Renegade Houses

RENEGADE HOUSES

Eric Hoffman

Illustration by Suzanne Clee

"This book is an anthem to the American spirit. It gives substance to words like 'inventive' and 'resourceful' and 'practical.' As more of us become less able to afford what the marketplace offers, Hoffman points the way to alternatives that are personal, practical, and affordable."
— *Robert L. Judd, Director of the Office of Appropriate Technology for the State of California*

Running Press
Philadelphia, Pennsylvania

Copyright © 1982 by Eric Hoffman
All rights reserved under the Pan-American
and International Copyright Conventions.
Printed in the United States of America.

Canadian representatives: John Wiley & Sons Canada, Ltd.
22 Worcester Road, Rexdale, Ontario M9W 1L1
International representatives: Kaiman & Polon, Inc.
2175 Lemoine Avenue, Fort Lee, New Jersey 07024

9 8 7 6 5 4 3 2 1
Digit on the right indicates the number of this printing.

Library of Congress Cataloging in Publication Data
Hoffman, Eric
Renegade houses.
Includes index.
1. House construction.   2. Dwellings.
I. Title.
TH4812.H63   1982   690'.837   82-14032
ISBN 0-89471-181-1 (paperback)
ISBN 0-89471-181-2 (library binding)

Cover design by Jim Wilson
Typography: Linotype Garamond by Duke & Company
Printed by Port City Press

This book may be ordered from the publisher. Please include
seventy-five cents postage. **But try your bookstore first.**
Running Press
125 South 22nd Street
Philadelphia, Pennsylvania 19103

Table of Contents

Dedication: To William and Louise Hoffman, my parents

Acknowledgments

First, Cecile Champagne, my indispensable, pugnacious wife who edited, typed, and took most of the photographs. Second, those who gave valuable assistance: Ted Adams, Diane Angelo (on whose work some of these drawings are based), Magnus Berglund, Jim Bierman, Doug and Pam Brouwer, Cecil Carnes, Mary Helen Chappell, Alfred Douglas, Rick Dyson, David Easton, John Forbes, Judy Gage, Mary Gordon, Tony Grant, Al and Clarice Johnsen, Larry Johnston, Bob Judd, Michael Leeds, Marvin Lichtner, Dick and Sherry McDermott, Kathy O'Neil, Jeanne Rosen, Paul Schurch, Wes Smith, Mark Sproul, Martha Sternberg, Patty Stumpf, Larry Tower, Tom Watson, and those who elected to remain anonymous. Third, many thanks to the Owner-Builder Center in Berkeley, California, for helping with the bibliography.

1. *Getting Started*

A HOUSE is the most important investment many Americans will ever make. If you don't already own a home right now, your chances of *ever* owning one are not great. With most existing loans unassumable, high interest rates, tight money, huge down payments, and an inflated housing market, a person making $40,000 a year can't qualify for a mortgage to buy an ordinary home in many parts of this country.

Renegade Houses explores innovative, economical approaches to solving your personal housing crisis. These successful examples of owner-built homes should offer you hope, inspiration, and direction. Not that all of them necessarily fall into the traditional concept of a life-long home. To make their dreams materialize, many of these owners viewed their creations as attractive practicalities, rather than as structures to meet someone else's expectations.

The people in this book approached the housing problem in a way that let them survive on their own terms. Some owner-builders had strong beliefs that traditional housing is wasteful and outdated. Others were driven by simple economic necessity. Often, a family needs two incomes to go on paying for a house, but the people in this book discovered alternatives to this kind of pre-planned existence. Whatever their reasons for building and adapting in unconventional ways, the end results were usually expressions of pragmatic individuality.

Building a house no bigger than needed is easier and cheaper to construct and maintain. Ingenious recycling of materials, money-saving construction techniques, working in new me-dia, clever interpretation of zoning ordinances, bargain purchasing, and built-in, energy-efficient features made home ownership a reality to people who would otherwise have been left out of the housing market. This book is the story of their inspirations, successes, discoveries, and mistakes.

Among these owner-builders, there is no stereotypical do-it-yourselfer. Their backgrounds, gender, professions, and levels of education vary. I met teachers, designers, carpenters, businesswomen, boat builders, counselors, farmers, professors, plumbers, musicians, artists, writers, government workers, attorneys, radio producers, engineers, photographers, and policemen who had only one thing in common: having created their own homes. Some were skilled in one or more areas of construction, but most were seriously lacking in training and experience and had undertaken building or extensive remodeling for the first time.

These do-it-themselfers often had difficulty accepting the idea of building their own living spaces. One, with a Masters degree, said, "I had no training and lacked the orientation. My life had been teaching English, not construction. The amount of skill and knowledge seemed overwhelming. I kept telling myself I couldn't do it." But ironically, you can acquire the techniques needed to construct an entire house by reading a couple of books, learning how to use about ten tools, and talking to knowledgeable people. As one owner-builder explained, "Building a house with little training is really a lot less daring than what our forefathers did. If they survived the covered wagon trip, they built a home with a saw and an axe. Really, we have it pretty easy." After conducting dozens of interviews, I realized that the only essential qualification is willpower. Builders who were able to overcome the "can't-do-it" mental block have now achieved financial gain, the satisfaction of seeing a house come together, and most of all, the dream of owning it.

You don't have to know everything about building houses. You just have to get through building *one* house *once*. It *is* important to approach the subject methodically and study each step before acting. There are countless books on the mechanics of building, the best of which I've included in the bibliography. Knowledgeable friends, neighbors, and relatives can be an indispensable source of help; invite them to the job site and pump them for advice. High school extensions, community colleges, and private organizations offer low-cost courses on different facets of construction. (A list of course names and locations appears in the Appendix.)

Building your own house is not a lifetime occupation. It usually takes from six months to two years, depending on what you're building and how you budget your time. Ed, a 35-year-old teacher, started building houses out of recycled materials during his summer vacations. He attributes his success to working adroitly with zoning ordinances and designing and building structures at extremely low cost. "After six years of college and twelve years of teaching I was making less than $20,000 annually. After four summer vacations of owner-building, I was worth about $250,000, and now I have rentals to pay my way."

Ed may be super-industrious, but everyone deserves a decent home, even if you have to build it yourself.

2. A Recycled Church

IN 1977, Doug Brouwer was driving through a rundown neighborhood in Gilroy, California, when an old, badly neglected church caught his eye. He left his car to look it over. A CONDEMNED sign was posted by the door. The belfry was leaning away from the rest of the structure. The paint was peeling, windows were broken, and graffiti was everywhere. Mounds of rubbish littered the interior. The stench of urine and liquor was overpowering. But Doug thought the church, built in 1870, was beautiful.

He contacted the owner who was planning to bulldoze the place to make room for a house. "It was such a beautiful structure, it had to be saved!" Doug quickly negotiated a deal: he'd pay the owner $1,500 for the structure and remove it within five months.

The back yard—the building was constructed between existing evergreens

Doug surveyed his new possession: "It was sitting on a redwood foundation that had settled around the perimeter. The church was all redwood except for the roof trusses and four major structural posts supporting the steeple, which were of fir. Termites had eaten the lower portions of the structural posts, which accounted for the lean. The redwood was ninety-five percent sound, and the trusses were in great shape. I wanted to move the church to our lot in a nearby coastal city."

His first major obstacle was getting a permit. Because the number of construction permits granted each year was limited, Doug applied for a house-*moving* permit. Doug had wanted to move the church in sections, but that was impossible. The church had to be totally dismantled, so technically, Doug would be building

Front door of the recycled church

from scratch. Nevertheless, he argued that he was moving an existing structure, not building one, and it worked.

As it turned out, a house-moving permit had other advantages: inspectors would not visit so often, and Doug avoided the expense of paying a lumber grader to verify the structural soundness of second-hand lumber.

Taking the church apart proved to be easier than Doug anticipated, since it had been built with square nails that tapered towards the tip. Once the head is pried out, a square nail loses its grip. By pulling the heads just a quarter inch, Doug could remove a twenty-foot piece of siding without splitting or damaging the wood.

"Former parishioners would come by and tell me about a marriage or funeral held in the church. They were grateful to know it was going to have a new home. Taking the church apart was an awful lot of work, but it was really interesting to see the old-time construction techniques."

The trusses were handmade and bolted together. Doug carefully marked each truss so it could be reassembled later, and took great care to stack the dismantled materials in a readily usable order. "I had the floor joists in one stack, the subflooring material in another, the framing in another, and so on. That way I didn't spend days going through stacks looking for what I needed."

He rebuilt the church on a ten-inch T foundation and converted the interior into a spacious two-story house. Because the anteroom did not add aesthetically to the structure, Doug left it off and used the materials to replace damaged wood elsewhere. The Building Department required an engineering report to verify that the original structure could meet today's Code. The only modification required was the addition of two plates on top of the 2″ x 6″ redwood stud walls. The studs varied in width from 5½″ to 6½″, so Doug and his wife Pam spent a few days ripping the oversized studs into uniform 5½″ widths. The tall Gothic windows were lowered about six inches so the arches wouldn't interfere with the placement of the second floor. "But anyone who had visited the church before wouldn't notice the windows had been lowered and moved laterally." The structural members under the steeple were replaced with new wood, and the trusses were assembled and pivoted into place with a block and tackle.

Before covering the framing, Doug decided to strip the paint from the sheathing and interior paneling. He built a plywood box four feet wide, twenty feet long, and one foot deep and lined it with a heavy rubber material to make it watertight. Each day, he stripped two loads of sheathing in a sodium hydroxide solution. Paint separated from the boards after 12 hours' soak. (To protect himself, Doug wore a mask against the fumes, a raincoat, and rubber gloves to avoid skin irritation.) In less than two weeks, all the sheathing was clean.

The roof was 1½" foam covered with ply-

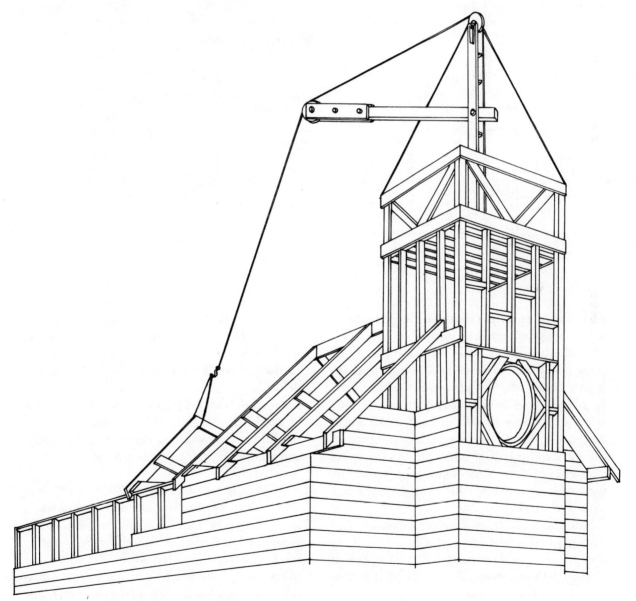

Using a block and tackle to rotate the reassembled trusses into place

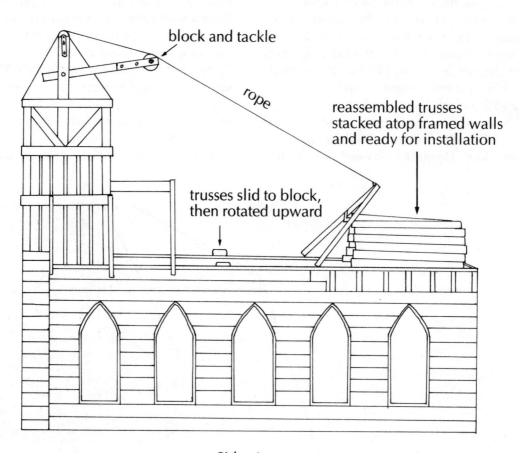

block and tackle

rope

reassembled trusses
stacked atop framed walls
and ready for installation

trusses slid to block,
then rotated upward

Side view

wood and composition shingles. Doug designed eight skylights utilizing "unsalable" tempered glass shower doors. "They have a slight green tinge which wasn't too popular. I bought them from a dealer who was glad to get rid of them. Each skylight—half a shower door and recycled wood for frames—cost about $5 to make. I also designed a skylight running the entire peak of the roof, comprised of thirty pieces of shower-door glass. The skylights were inexpensive, but they took an awfully long time to build and make watertight."

Plumbing is copper pipe for incoming and ABS plastic for outflow. But if he had to do it over, Doug would use cast iron: "Plastic pipe is loud in the walls, even when wrapped in insulation. It's also a fire conductor."

Framing materials for the interior walls were mostly recycled studs from the original anteroom. The joist system for the second floor, subfloor for the second floor, and the finished hardwood maple floor were all salvaged from a demolished college dormitory. Doug bought the materials for only $50 on the condition that he do the dismantling. It took him a month of careful crowbar work and endless nail pulling, and another month to ready the maple floor for installation.

The house has more than 2,700 square feet of floor space. Downstairs are four bedrooms, a large entry area, bathroom, and sun room. The walls are finished with expertly taped, untextured sheetrock, painted white. (The fact that the walls did not need texturing to cover mistakes testifies to Doug's perfectionism.) The hardwood floors, solid oak doors, and oak molding give the house an impeccable finish. Upstairs, Doug installed a spacious kitchen, study, bathroom, and "big room," finishing with the original redwood paneling and molding. Doug and Pam did not compromise their commitment to economy and recycled materials. The redwood kitchen cabinets came from a bankrupt restaurant. Doug took a welding class and later welded into place a carefully designed staircase, with a redwood bannister he carved himself.

Even though their son was born during the course of their work, Pam performed many steps of the construction and was personally responsible for some of the finer finishing touches: "We couldn't afford to buy this many stained glass windows, so I took a class and made them. I prefer looking at them to making them, though. They were tremendously time-consuming." Pam also laid most of the finished maple floor.

Closeup of the truss system and peak skylights

Iron chandelier and translucent skylight

When asked about costs, she rolls her eyes: "We've saved plenty, but on the other hand, we've given up a tremendous amount of time. This project has dominated our lives, which to me is part of the cost. But we keep telling each other it's worth it."

Doug smiles and answers in monetary terms: "It's hard to figure exactly. Basically for the three years while we've been building, we've been living on $55,000 in the form of a family loan. That covered building and living costs. The project feels like about seventy-five percent labor and twenty-five percent materials."

Are Doug and Pam planning other building projects? "Nope! We're going to sit back and enjoy each other, our son, and our beautiful home."

The first-floor bathroom

The kitchen, looking out on the second-story porch

Circular staircase in the foyer

Another view of the
kitchen, showing the shower-door skylights

17

3. Home Was a Water Tower

MICHAEL LEWIS, designer and artist, doesn't like paying exorbitant rents. "An apartment leaves no room for creative expression and individualism. But a house is too big for one person, so you have the hassle of sharing it with someone."

Years ago, Michael spotted a large unused water tower on a friend's property and somehow envisioned it as home. For $25 a month rent, the water tower was his. It took three months and $2,500 to convert the dilapidated water tower into a beautiful, innovative dwelling.

The circular, redwood water tank (20 feet in diameter) became a bedroom, with windows added for ventilation and light. To get to the bedroom, he built a handcrafted staircase out of 8" x 12" redwood. Michael claims the idea came from a hand fan: "A two-inch hole was drilled in each sculpted step, which was then slid onto a two-inch pipe serving as the stairway's center post. Once the post was in place, I fanned out the steps and fixed them in place with shear pins."

The 14' x 16' area directly under the water-

The original water tower is still visible at left

tank bedroom became the kitchen and entryway. To get enough headroom, Michael sculpted 4½″ off the ceiling joists to create a nicely arched canopy. To finish the kitchen ceiling, he nailed tightly-strung mesh to the inside of the joist and plastered over it.

It may seem the height of folly to carve chunks out of a joist system that holds up the floor or ceiling. "But the joist system below a water tank is always over-built because it's designed to support tons of water. With the tank empty, I felt it would be safe to shave down four of the joists to get enough headroom in the kitchen for a six-foot person."

Off the kitchen, Michael added a 10′ x 12′ dining room. A slab foundation was poured to butt against the existing slab that supported the water tower. Conventional framing, large salvaged windows, and a pitched roof with redwood shingles completed the dining area. The new and old concrete slabs were covered with grouted, one-foot-square quarry tiles, drawing the kichen and entry/dining area together.

The kitchen tile includes many pieces of broken pottery—an idea that came to Michael when his cat knocked a stack of dishes to the floor. He liked the patterns on the shards so much he ground down the sharp edges and included them in the tile countertop. When he needed curved pieces for the coping around the sink, Michael carefully cut up some of his favorite coffee cups.

Michael says the keys to inexpensively and ingeniously renovating a structure are being innovative, patient, opportunistic, and constantly on the lookout for salvageable material. "There are so many ways to get good second-hand material if you're willing to look over a period of time. Before starting to build, a person should go to garage sales, salvage yards, and watch the want ads. Often what you find will

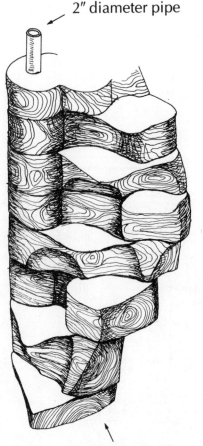

2″ diameter pipe

individual redwood stairs

View from top of stairway

give you inspiration for an entirely new idea."

Following this philosophy, Michael found an old gas range for $75. Fifty dollars more made it fully functional and beautiful. Bought for $60 at a garage sale, the kitchen cabinets were stripped, sanded, and stained. Michael also bought all of the windows, most of the wood, all essential plumbing fixtures, and an antique free-standing German fireplace for no more than half price.

A small bathroom built onto the side of the kitchen contains a sink, toilet, and easy-to-install fiberglass shower. The toilet discharges into a 700-gallon redwood septic tank, while the other drains are directed to a nearby orchard. The incoming water pipes are plastic.

"People keep telling me plastic pipe won't take hot water. All I can say is, after seven years there are no leaks."

To one side of the water tower Michael added a 20′ x 16′ living room on a pier-and-post foundation. Flooring is 2″ x 6″ kiln-dried pine which won't shrink. Normally used for decking on exposed-beam ceilings, this material makes an attractive finished floor as well as the subfloor—a great money saver. Michael turned the beveled sides down so there would be no grooves to collect dirt.

The inside walls are covered with inexpensive plywood, but Michael doesn't believe their plainness detracts from the tile floors and beam ceilings. "This room's interior walls are the

Side view, showing the entryway and dining area

cheapest ¼″ plywood on the market, but that's not what you see. Walls should be covered with art—hangings, artifacts, paintings, driftwood, or whatever. Art should be your visual wall, not overly-expensive siding or wallpaper that is later covered with art. The plywood just holds in the insulation!"

Michael's advice to other builders: "Don't listen to what other people say you need. Build what *you* need. Also, a little planning before saves a lot of money later. If you stockpile bargains before you start building, you won't feel pressured to buy new materials to get the job done."

Left: One of the recycled caryatid/faces that flank the dining area

The dining area

Closeup of the front porch and wind chimes

Below: Bedroom window with pediment (note rippled glass panes)

4.   *Too Steep and Too Old—They Said*

To PALO Alto teacher Larry Johnston, 1977 looked like a dismal year. "That was when the average Mid-Peninsula home became worth $100,000. I knew my meager $17,000 salary couldn't open any doors in the local housing market."

Larry watched the newspapers and listened for the faintest rumor that might lead him to that elusive fixer-upper, or "handyman special," priced at half the cost of other homes. After months of vigilance, nothing materialized. Then one day, a colleague told him about a small, shack-like structure whose elderly occupants had moved to a rest home.

Larry investigated and found a one-bedroom, cabin-sized structure leaning about ten degrees

"Too steep" house seen from bottom of the slope

The interior living/sitting room

and sitting precariously on a 45-degree slope. The roof had a noteworthy swayback. The realtor who appraised the house for Larry felt it was worthless and recommended the purchase be directed towards the value of the parcel which, after all, was situated in a wealthy hillside suburb.

Sitting as it did among large expensive homes, the little house looked out of place. Built in 1926, it had been the guest house on a large estate that had been the summer retreat of a wealthy San Franciscan who, among other things, was a boxing promoter. His boxers stayed in the guest house and trained in a ring attached to the rear of the structure. (When Larry first visited, the ring was still there.)

Apparently the 1930s promoter understood the overall trend in Bay Area real estate. He had the property on which the guest house sat recorded as a separate parcel, even though it was maintained as part of his estate. The old couple who lived there had been leasing for decades; when they left, the promoter's heirs wanted to sell. Larry offered $48,000 for the tired little place and its near-vertical lot. A deal was struck at $50,000.

Larry first worked to save the house. "Its stone and wood foundation had slowly crept down the hill, causing the house to lean. I paid a guy $3,500 to jack up the house and put a foundation under the front part." Even though Larry couldn't afford an entire foundation, the newly replaced portion was soundly made. A series of six-foot shafts tied the foundation into the bedrock. The rebar in the shafts tied into the rebar in the above-ground foundation. To ensure the greatest strength, the shafts and foundation's conventional portion were poured at the same time.

Even the house's framing needed reinforcing.

"The house didn't even remotely conform to building codes. In some of the walls, studs were 30 inches apart." Larry tore off a great deal of siding and added studs and insulation. To restore a straight roofline, he doubled the number of trusses under the roof. He had to fumigate to rid the house of termites, tear out rotten underpinnings, and replace most of the windows. "For around $9,000 I made the house sturdy enough to survive into the foreseeable future."

An avid gardener, Larry wanted to make his steep lot useable. The lot was so steep that a person standing six feet behind the house could look down on the roof. "To make their land productive, the Incas built terraces on hillsides steeper than mine." With pick and shovel and a great deal of backbreaking work, Larry terraced his property. Behind the house, he dug out a walkway three-feet-wide and poured a six-foot retaining wall. He dismantled the boxing ring, carted it to the city dump, and leveled the backyard against the retaining wall.

Stairs lead to a special wooden platform for plants

View of the roof, from the top of the slope

The roof is only three feet from the retaining wall and at the same level, so Larry connected them with a ramp: "The roof received the most sunlight on the property and I just couldn't see passing up all the space. From underneath, I reinforced the roof some more and designed some planters that fit the roof's slope. I laid outdoor carpeting over the roof tiles in areas that would receive foot traffic. In the winter, when the asphalt tiles are brittle, I don't allow people onto the roof. I wouldn't recommend walking on a roof for others. We're very careful." He even modified his cast-iron vent pipes to support potted plants; today the roof is a beautiful array of flowers and vegetables.

Larry decided to take his terracing skills indoors and "decided to make an extra room under my house." With a pick, shovel, and wheelbarrow, he hauled tons of soil until he had created a 12′ x 18′ flat space underneath the house. He poured a concrete slab and retaining wall,

Stairs of brick and mortar

my advice is to rent to a carpenter, who wants to trade rent for services." As for what it takes to make a fixer-upper into a viable home: "First there was luck and timing in finding a place, then there was taking a chance. The big variable was working hard; moving dirt and staying up nights until it was finished. It was a rundown cliff dwelling, now it's a cozy hideaway with attractive gardens. The whole process was good therapy."

framed in the walls, sheathed them with plywood, cut in a door and window, wired in some sockets, and rented the room.

John, Larry's first renter, was a carpenter who offered his services for reduced rent. "Naturally, I supplied the materials. His knowledge of how a house is put together and his skills ushered in major improvements: 250 square feet of redwood deck, a remodeled kitchen, brick walkways, and repaired plumbing. It was great to leave for work and come home to improvements." During summer vacation, Larry went to Alaska and returned to find a surprise: "John had torn out the wall between the kitchen and front room which made the tiny rooms grow into a large room. He sheathed the front room with grapestakes because they're attractive but inexpensive."

John stayed about a year. Larry: "Of course,

5. *Warehouse Conversion*

IZZIE LIEBOWITZ sees things other people don't see—like Michelangelo seeing David in an unsculpted block of marble. An artist, designer, and inventor whose work has appeared in national publications, Izzie has designed prestigious restaurants and bars in the Midwest and on the West Coast. His innovations in opalescent art glass design have been compared to those of Louis Tiffany.

Izzie's unique qualities don't readily transfer to other mortals. However, his attitude towards housing, and the way he goes about creating a living space, *are* transmissible: "There *is* no housing problem. Anything with a roof over it can be made into an attractive home. The possibilities are unlimited."

Izzie has lived in an assortment of structures, none of which would be classified as conventional. He takes pride in creating a home where others don't see one. In 1979 he discovered the

Sundeck opening off studio at rear of warehouse

oldest surviving foundry in California.

Built in 1847, the predominantly redwood structure sat in a downtown area. From the front, the wooden facade (which needed painting) seemed to belong in a set for a Hollywood Western. The rest of the building was covered in corrugated metal. During the second half of this century, it had served as a warehouse, but was now at the to-be-or-not-to-be point all buildings eventually reach. "To be" meant repairs and further investment. And before Izzie invested his $52,000, he discovered that the building was zoned for nonresidential commercial use. "I put a contingency on the purchase: before the deal could close, the zoning would have to be changed to include single-family residential use."

Izzie applied for a zoning variance. "This was the most critical point in making the structure a home. I presented the Zoning Department with

One of Izzie's vintage metal signs

a complete set of plans that would upgrade the structure's appearance and use. I also spelled out the benefits of revitalizing California's oldest foundry, pointing out that the Chamber of Commerce likes to boast about attractive historical buildings. Preserving a building with such

A portion of the downstairs work area

28

a rich history also drew some support from the local historical society." The variance was granted.

The old foundry sat on its original redwood foundation. "A 135-year-old wood foundation made me a little nervous. After all, wood resting on soil is supposed to rot or be eaten by bugs, so I had the foundation and framing checked. There was some dry rot, but, to my surprise, no termites. The tight-grained, original-growth redwood used in that foundation can last a lot longer than the second- and third-growth redwood they harvest nowadays."

The two-story building has over 6,300 square feet of floor space. Izzie's plans called for making the 4,900-square-foot ground floor into a workspace and the 1,400-square-foot second floor into living space. "But before I could get started on the inside, I had to make sure I could keep out the weather. The roof leaked, and the window frames were badly rotted, and missing panes needed replacing. I replaced the leaky corrugated metal sections with new material and replaced the windows."

For working on the second-story windows, Izzie designed what he calls a "gravity-clamp scaffold," resembling a short diving board. Two sections fit snugly against the inside and outside wall of the windows. To them is bolted a platform that extends out into space. The weight of the worker standing on the platform outside the window pulls the two parallel sections snugly against the wall. Izzie cautions that the structural soundness of the wall beneath is all-important because, "If the wall gives way from the person's weight, down goes scaffolding, person and all." To play it safe, Izzie wore a safety rope. "That platform made replacing the windows easy. Instead of climbing up and down a twenty-five-foot ladder with tools and material, I could stand on my platform while my girl-

Insulation being installed during renovation

The kitchen ceiling

29

friend inside handed me tools and materials."

After replacing the windows, he insulated the second floor's ceilings and walls. "Because of the expense involved, I elected to insulate only the living area." Then he prepared the second floor for habitation: "I repaired the stairs and replaced a few weak floorboards. Following the plans, I chalked in the areas where the interior walls would be." Using standard 2″ x 4″ construction, Izzie framed in a living room, bedroom, kitchen, and bathroom.

Next, he plumbed the second floor and expanded the existing wiring to accomodate his needs. "The warehouse already had a downstairs bathroom connected to the city's sewer system, so even though I had to replumb the downstairs and plumb the upstairs, I wasn't faced with the red tape and expense of a sewage hook-up." Installing the plumbing wasn't difficult because the exposed framing made access easy.

Closeup of the thermostat-controlled water mixer

The bathroom, as seen from the bedroom

With the wiring and plumbing in place, Izzie started finishing the rooms. In the kitchen, he created a steep-peaked, open-beam ceiling. Typical of Izzie's approach, he utilized what existed to help create what he wanted. The proposed rectangular kitchen lay against an outside wall. The rafters didn't peak until they passed over the kitchen and reached the bedroom and bathroom. But using pine, Izzie sheathed the rafters to the point midway above the kitchen. At the kitchen's midline, he nailed into the original rafters a second set of "cosmetic" rafters that descended to a new wall parallel

to the outside wall. Izzie finished sheathing the old and new rafters with pine, enclosing the ceiling.

The sink is set in a discarded pool table "island" in the center of the room. Since the slate had already been sold, Izzie was able to buy the table for only $50. The drainboard is made of 1" x 1' x 1' quarry tile. Izzie bought the 1920s gas range from a junk dealer for $150. He found the cabinets offered for $150 in a newspaper want ad. "People are always remodeling their kitchens, so it pays to keep watching the want ads."

In the bathroom, Izzie plumbed a freestand-

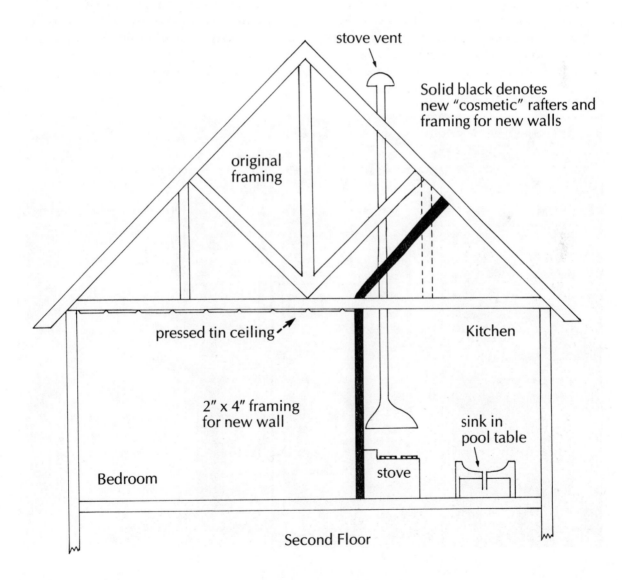

stove vent

Solid black denotes new "cosmetic" rafters and framing for new walls

original framing

pressed tin ceiling

Kitchen

2" x 4" framing for new wall

sink in pool table

stove

Bedroom

Second Floor

31

ing, cast-iron tub and enclosed it in lathing material. Attached to the bathtub spigot assembly is a thermostat-blender from a photographic laboratory. The desired water temperature is achieved by setting the dial on the blender. "I don't know why more people haven't installed something similiar. There's no fiddling with faucets, and the water doesn't flow until it's the right temperature."

The toilet is the conventional $50 porcelain variety. But Izzie detached the water closet, put it high on the wall, and connected it to the bowl with copper pipe. Now the toilet looks like the old-fashioned $200 variety.

The bedroom and living room boast an ornate ceiling that looks as if it were lifted out of Versailles—or at least from a very fancy, nineteenth-century Victorian mansion. What appears to be carefully sculpted plaster is actually pressed tin. The W. F. Norman Company in Missouri embosses patterns into sections of tin that when joined together, make an ornate surface. Commonly, the material comes in 2′ x 2′ squares. "The metal is easy to install. I just laid a light joist system, 2′ on center, across each room and nailed blocking every two feet, creating a matrix of two-by-two squares. To each square I tacked a metal sheet. It took two days to cover 800 square feet of ceiling." Around the perimeter, Izzie added a pressed-tin cornice and painted the whole ceiling a soft white.

In the living room and bedroom, he covered the walls with sheetrock and hired a professional to tape and texture them. He covered the floors with wall-to-wall carpeting. All the doors are oak, and antique oak furniture complements them.

In the work area downstairs, Izzie painted the

Pressed tin ceiling, after painting

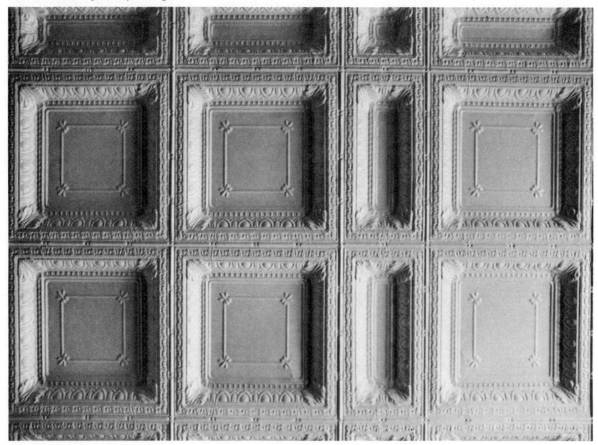

Cornice and molding of the bedroom door

old water-stained wood walls. His original concept of renting the cavern-sized area never materialized. "As it turned out, I needed a lot of space for different design jobs I was doing. Once in a while I rent some space to another designer for short-term projects." In the corner sits a beautifully designed stained glass window, ready for shipment to a wealthy client. Hanging from the ceiling is a vintage model of a Flash Gordon rocket ship, complete with lights and moving parts. It will be shipped to a bar featuring a 1950s concept of outer space—a theme developed by Izzie.

At the back of the warehouse, Izzie laid a joist system through the original horizontal trusses and let the pitched roof form the walls and ceiling for a 300-square-foot elevated studio. This triangular room is deliberately hidden away from the structure's activity areas: "I wanted a quiet place where I could draw and plan." The studio is sheathed in mahogany. "With the slowdown in the housing and boating industries, there's a glut of mahogany on the market. I bought oddly-colored and warped wood from an importer for 15¢ a board foot and had the wood milled into sheathing material."

After sheathing the ceiling, Izzie decided the room needed more light. He tore out the end wall and reframed it with plenty of glass. In the center he installed a small mahogany door that was once on an old tugboat. The door opens onto a redwood deck that Izzie suspended from the adjoining building. To get to the loft, Izzie built a lightweight staircase out of ⅝" plywood and 2" x 4"s. "I enjoy designing staircases. I've made about ten different models."

As Izzie sees it, the key to the housing problem is attitude. "This country has so much going on; there's so much sitting around waiting to be used. Too often, people won't touch a product unless someone tells them it's okay. This stifles the creative spirit we all have. People should think of something's utility first and then think of incorporating the utility aesthetically. It's fun and challenging to think of housing in this context. As long as you let someone *else* define what looks good, then you're paying for status, and you'll pay through the nose. By definition, status will cost you. That's what status is all about."

6.   An Inner-City Hideaway

FOR FOUR years, former *Life* photographer Marvin Lichtner had lived in a luxurious sprawling home in Beverly Hills. His friends couldn't figure out why Marvin would rather live in a rundown inner-city neighborhood. But Marvin had his reasons: "I grew up in the Bronx and always loved the pulse and smells of a city. More than anything, I'd grown tired of the conspicuous-flash lifestyle and the inconvenience of being forced to jump into a car for a ten-minute drive every time I needed cigarettes. I wanted to walk to a corner market. But also, I knew I could sell the Beverly Hills house, buy an inner-city place, and have plenty of money left over."

Still, Marvin had problems communicating his wants to realtors. "I told them I wanted an older house in a neighborhood with a healthy ethnic and cultural mix and a little graffiti on

Top floor of the carriage house, as it originally appeared

34

the sidewalk. They kept showing me restored Victorians in middle-class neighborhoods—houses they apparently thought I ought to buy."

Taking matters into his own hands, Marvin watched newspaper ads and drove around neighborhoods until he found a tired, vacant Victorian. At the rear of its double-sized lot sat a turn-of-the-century carriage house that appeared to have been derelict for decades. But Marvin like what he saw and checked the Assessor's records to find who owned it. The owners, a couple embroiled in a contested divorce, were both leaving the state and wanted to sell, but hadn't gotten around to listing their house. "I offered them a large cash down payment and got a good deal."

Marvin developed a plan for his new property: "I intended to maximize the use of both

Ship's ladder leads up to study/library

Kitchen of the attic apartment

structures. Even though the property was zoned single-family residential, I saw three, and possibly four, separate living spaces. The house, of course, was one. The attic could easily convert into an apartment, and the two-story carriage house, with its 1,700 square feet of floor space, could become another two units."

Marvin enclosed most of the property before taking the steps to transform it into a well-landscaped private garden. "I erected a ten-foot fence from the side of my house to the neighbors' fence, installed an electronically controlled door and intercom which made the yard inaccessible from the street." Marvin also added two feet to the existing 6′ fences separating his property from the neighbors' side-yards.

Marvin began by renovating his main house. He remodeled the kitchen, put in a new freestanding fireplace, replaced the kitchen and bathroom linoleum with tile floors, patched and painted the walls, and installed wall-to-wall carpeting.

Next, he began converting the 1,000 square foot attic to a living area. Typical of most turn-of-the-century architecture, the Victorian had a steeply-pitched roof. From the roof trusses, Marvin removed the cross members and braces and replaced them with support posts carefully placed over bearing walls. "The restructuring made the attic a walkable space."

To brighten up the attic, Marvin cut through the roof and installed several skylights fitted with clear glass. The ceiling joists were strong enough to support a 1″ fir floor. But before Marvin laid it down, he worked the plumbing through the joists. He added a sub-panel to his main circuit panel, wired three circuits into the attic, and insulated. The walls and ceiling are mostly sheetrock. But because parts of the rafter system had complicated configurations, Marvin hired a professional to apply lathing and plaster over the irregularly shaped areas. As a result, the finished ceiling has nicely flowing curves that couldn't be achieved with sheetrock alone. He made the bathroom and kitchen cabinets and artistically tiled them. To provide private access to the attic apartment, Marvin built an outside stairway.

Marvin rented the house, moved into the attic apartment, and began working on the carriage house, "which was a much bigger undertaking than the attic and the house combined. Essentially, it was a framed building with a mostly rotten wood foundation and no plumbing or electricity." Marvin hired a contractor to jack up the building and pour a concrete foundation under it.

The window frames were rotten, and many of the panes had fallen out. The roof was a combination of sixty-year-old shingles and tarpaper.

Another view of the former attic

36

Bush, painted black, silhouetted against a white wall

Marvin tore out the windows and stripped the roof. Many of the trusses, doors, and most of the interior wood were covered with old, cracked paint, so Marvin hired a sandblaster to clean them up. "He blasted the paint and water stains from the trusses, the 1" x 6" ceiling boards the shingles had been nailed to, and the doors. He didn't even take the doors off their hinges. For $250 and one day, it was a super deal." However, Marvin warns that a house should be sandblasted *before* one moves in. "A friend brought in a sandblaster to remove paint from a fireplace in a fully-furnished house. He covered the entire front room in canvas, but by the time he finished, he had to throw out his rugs. Months later he was still finding pockets of grit throughout the house."

In the ceiling, Marvin installed two 4' x 4' double-paned skylights. Usually such skylights cost $800, but he got his for $200 each from a contractor whose client had rejected their slightly discolored glass. To finish the roof, Marvin laid ½" plywood on the original 1" x 6", put 1" rigid foam on top of that, and tacked on a shingle roof. He installed new custom made wooden windows, making the carriage house waterproof for the first time in more than forty years.

The upstairs floor was in poor condition, so Marvin sheathed over it with #3 kiln-dried fir (the cheapest nonshrinking wood he could find) and later sealed it with oil-based Varathane. Marvin wired and plumbed the carriage house, insulated the walls and covered them with sheetrock or barn wood. Up in the trusses, Marvin added 300 square feet of floor space by building a loft—accessible by a ship's ladder—that serves as a small library and study. "When I

Knicknacks among the trusses

weighted it down with books, one truss sagged a little so I put support posts under some of the trusses."

Throughout his reclamation project Marvin worked with the city's building department to obtain all the necessary permits. "They had no objections to my saving the old carriage house, but they wouldn't let me put in a kitchen." So before putting in a sink and range, he bid the city farewell: "I finished everything *but* the kitchen and called for the final inspection. The inspector knew what I was doing and even offered friendly advice. He said to wait four

months in case his boss assigned him to reinspect. I heeded his advice."

Next Marvin spruced up the enclosed grounds. He laid thousands of bricks, trimmed existing trees and planted tree ferns, fuchsias, rhododendrons, and other colorful plants. Today his renters call the grounds "Marvin's Gardens." Lying in a hammock, it's hard to believe you're in the middle of a city. It's quiet and peaceful. Birds chirp, greenery is everywhere, and the only backdrop is the weathered siding on the carriage house.

7. The "Progressive" Depot

AS AIRPLANES and automobiles became popular, railroad lines drastically cut services on their passenger routes. Left behind were thousands of train depots designed by leading architects of the Victorian Age. The railroads tore them down or sold them for $1.00 apiece.

In 1958, Southern Pacific terminated the Suntan Special which had linked the San Francisco Area to beach resort communities along Monterey Bay. After the line closed, a Southern Pacific executive purchased the Capitola Depot. The depot's walled-in area measured 22' x 50', containing a raised baggage room, telegraph office, centrally-placed ticket area, and waiting room with lavatory. In keeping with its turn-of-the century architectural style, the depot

The Capitola railroad station, as it originally appeared . . .

. . . And after renovation

boasted a thirteen-foot ceiling. The executive planned to move it to a nearby bluff and make it his summer vacation home. But the civic fathers turned down his plan, on the grounds that the depot would be out of place among the staid Victorians and the wealthy summer houses already occupying the bluff.

For $1.00 the rebuffed executive sold the depot to Lucina Savoy, a designer and antique dealer who recognized the old structure's great potential. She thought the building's front, with its classical Doric colonnade and semi-circular ticket area, would combine naturally with a French classical antique motif. Further, Lucina saw nothing odd about combining her sophisticated tastes with the raw tenacity it takes to pound nails, lay brick, and replumb an old sink. In transforming the structure into a

home, Lucina stuck to each task until it was finished. If she wasn't satisfied with her work, she'd tear it out and do it over.

Without compromising the building's lines, Lucina added 400 square feet of floor space by walling in the porticos at both ends of the building. She expertly faced one of the porticos with antique bricks, leaving the columns exposed. She poured a conventional T-foundation, laid a subfloor, and attached the framing to existing eaves. The end rooms, with their exterior doors, were not opened to the rest of the house, but were used for storage.

After spending a great deal of time sanding and stripping the green and yellow paint off the columns and door and window trim, Lucina painted them white. She then covered the entire exterior—excluding the brick wall—with cedar

40

Off the front porch, a carport and garden

One of the original pillars and wooden cornice

shingles and lightened them with an application of diluted bleach. In combination with the extensive white colonnade and trim, the shingles create an elegant exterior.

Inside, Lucina remodeled extensively. Rooms were assigned new uses: the raised baggage area became the bedroom, the telegraph office became the kitchen. The ticketing area with its semicircular wall became the dining area, and the waiting room became the entry way and front room.

Throughout the interior, Lucina installed wainscoting on the walls. Onto the remaining wall space and the ceiling, she tediously applied lathing and plaster to create a new surface and coved ceiling. She painted the bedroom robin's egg blue and the other rooms a pale terra cotta.

To make sure that the new multi-paneled closet doors matched the depot's existing doors, Lucina hired a cabinetmaker. She took the same

The Baggage Room door retains its original plaque

into a compact modern kitchen with a slate-look, plastic laminate counter, Jenn-Aire range, and dishwasher.

In the kitchen and entry, Harry pulled up the hardwood floors and laid tile, which added color and was easier to clean. In the corner of the dining area, he installed a fireplace encased in the wainscoting that appears on all the walls.

In 1971, Harry wanted to move to a beach house. He met realtor and dress designer Cecil Carnes who wanted the depot—and just happened to own a beach house. They traded structures.

Cecil appreciated her predecessor's efforts and good taste: "It just needed some finishing touches to become exquisite." She opened the wall into the remaining enclosed portico and made it into a study. She had the end wall torn out, inserted an 8' x 8', 32-pane window, and

The compact kitchen

care in making kitchen cabinets and a new bathroom vanity. The depot's waiting room had a lavatory, but no shower or bath. Lucina plumbed a combination tub/shower into the bedroom and built a closet around the tub. When the tub's in use, the doors are open; at other times they remain closed, hiding the tub from view.

Just when Lucina was about to have the depot totally transformed, she found herself heading for career advancement and a better life in New York. She sold the depot to Harry Schultz.

A practical do-it-yourselfer, Harry immediately got to work finishing the structure in Lucina's style. He opened the wall from the bedroom into the enclosed portico and converted it into a large bathroom complete with a sunken tub. He transformed Lucina's "kitchen"—a hot plate, sink, and cabinets—

The living room and stairs leading to the study

outside the window she fashioned an aviary populated with white doves. She laid oak flooring in the study and on the steps descending into the living room, carefully matching it to the oak flooring in the living room. She replaced the solid wood exterior doors leading from the living room with ten-foot-high French doors. Outside, she enhanced Lucina's original fuchsia and rose gardens with a number of rare rhododendrons. In the rear, she put in a 10' x 20' reflecting pool, complete with French statues. Through an importer who specialized in European antiques, Cecil found two eight-foot wrought iron gates and installed them in the driveway leading to the rear of the house.

Four years ago, she was asked if the depot's beautiful grounds could be used for a wedding. Cecil was delighted, and since then, several weddings have taken place at the depot.

Recently she has made the elegant robin's-egg-blue bedroom available for bed and breakfast. Proudly she points to a recent newspaper article that places her depot among the top five bed-and-breakfast establishments in the Monterey Bay area: "The depot has grown on me over the years, and I'm having so much fun with the bed and breakfast concept that I have my eye on two more out-of-use depots north of here!"

8. Land Leasing

AL JOHNSEN wanted to live in the country, but still be close to his work at the University. He wouldn't accept the idea of a long-term mortgage, or of paying high rent to a landlord.

So how did Al and his wife Clarice end up in a nice little house in the middle of a lush meadow, bordered by a stream, surrounded by ever-green and deciduous trees—all for an initial $5,000 down and $100 a month?

Al declares, "It's crazy the way so many people tie up all their money to get in on the American dream of home ownership. Plenty of hardworking husbands and wives can't afford a house even by pooling their paychecks. Today,

A simple rectangular dwelling on rented land

44

home ownership—assuming you can get in in the first place—takes a superhuman effort to sustain.

"Rather than become a slave to a mortgage or a landlord, I looked into alternatives. I knew the government used to lease land for cabin sites and the like. Would that idea work in the private sector? We got in the car and drove around looking for the ideal spot within a fifteen-mile radius of the University. We fell in love with this place."

Al went to the county government building and started asking questions. The Assessor's office said that "his" place was part of an undeveloped 270-acre parcel that had been owned by the same family for over a hundred years. The Zoning Department informed him the land was zoned for 20-acre minimum parcels. The smallest legal parcel — even if it were for sale — would be too large and too expensive.

By asking around, Al found out that the family had strong feelings about preserving the land's natural beauty and had resisted developers' offers to subdivide the property. So Al made the owners a proposal: "For fifteen years, I would lease five acres at the going grazing-rights rate. I would build a house acceptable to the property owner, deed it to him when I died or moved, and pay taxes on it in the meantime."

The owner accepted Al's offer. "In 1965, I leased the land for thirty dollars a month, and it cost me $5,000 in materials to build a 750-square foot house [*At today's prices, the cost would be closer to $15,000*]. A lot of people thought I was crazy to build a house for another man. But as time passed, the plusses greatly outweighed the minuses. Later, I felt guilty about paying only thirty dollars, so I phoned the owner and convinced him to take fifty dollars a month. That lease ran until 1980. I renegotiated for another ten years at one hundred a month."

Al wanted only ten more years, because he figures that may be all he needs. "Heck, I'm sixty years old. I might get an ingrown toenail and die from an infection. You never know."

Built with all the necessary permits, Al's house is functional, with an intelligently thought-out floorplan. In designing it, Al kept three goals in mind: "Keep the cost low, make good use of floor space, and enhance the environment." Most of the materials were purchased new. The framing is conventional stud wall with a beam ceiling. The living and dining rooms overlook the stream, and a large bedroom, kitchen, and bathroom complete the house.

Al later added a potter's studio, a sauna near the creek, and a guest house. He plans carefully and likes to work with uncomplicated rectangular floorplans. "That's why it took my son and me only one week to build the guest house."

Al figures he's increased the owner's equity by at least $100,000. "I'm glad to have helped him. But more important, look what I've done for myself! For $5,000 out of pocket and an incredibly low monthly lease, we live on one of the prettiest pieces of land in the area. My wife and I realize our two grown sons won't inherit, but we couldn't have bought it anyway. We've all been able to enjoy it here for next to nothing. We've been able to travel—and not worry about a mortgage in lean times. I can't think of a better arrangement!"

9. *The Umbrella Plan*

As COLLEGE buddies, Rick and Larry talked about buying land and living a simpler agrarian lifestyle. But by the time they graduated, nothing had come of their talk, and they went their separate ways.

Two years later, Larry was working as a carpenter when he heard about some parcels for sale on one of the rugged ridges overlooking the Los Angeles basin, and invited Rick to look over the property with him. They liked one ten-acre parcel priced at $12,000. Even in 1971, the price seemed pretty high for land with no access road, no electricity, and so densely covered with chaparral that they were forced to chop their way around the property to get a look at it.

Nevertheless, Larry and Rick thought the plusses outweighed the minuses. The property had a year-round spring—a rarity in the

Top of brick and stone fireplace, showing roofbeams

Larry's adobe house

parched Southern California mountains—as well as a large pond suitable for swimming, and a view of Los Angeles and the coastline.

As soon as the property was theirs, they had a dirt road cut in along the easement. During the first winter, they cleared brush: "We used chain saws and brush axes and pulled out a lot of stumps. Most of these semi-desert plants have tremendous regenerating powers. We brought in several loads of good soil for gardening and ten tons of sand to make a beach at the pond's edge. By the end of the first winter, we had cleared enough to get around the property, but it took us five years to get as fire-safe as we are now."

Larry and Rick considered building a communal structure and various outbuildings. But "when we sat down to draw a house plan, it became clear we wanted different things. We re-

spected each other and thought the best course was to develop a plan that would suit each of us." But the property, zoned for single-family residential and agricultural use, was not legally dividable. "We decided to divide the land informally between us and build two separate structures." Once acceptable boundaries were established, they developed the site under what Larry calls the "Umbrella Plan."

Rick took the more conspicuous building site and applied for a building site and applied for a building permit to legitimatize the single-family-residential zoning and satisfy the county that activities on the parcel were related to a legally-sanctioned house. (Sometime later, Larry would build his own structure, small enough to be considered a guest house.)

For a 1,000-square-foot octagonal house featuring a a massive brick and stone fireplace,

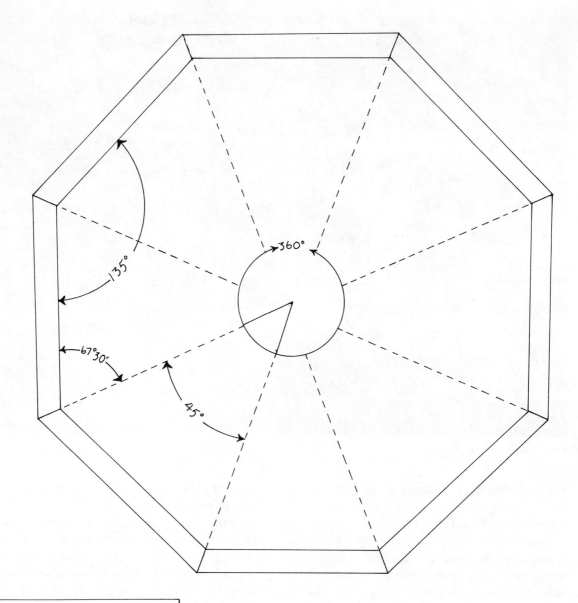

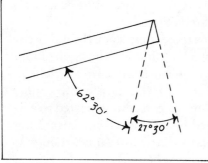

How to Make a Geometric Floor Plan

Basic formula: Divide 180° by the number of walls de-sired. The result is the angle at which the square ends of your board lumber must be cut. For the octagon shown here, 180° divided by 8 equals 27° 30′ of arc. (Cuts on the same board should *not* be parallel, but point toward the center of the polygon.) When two adjacent boards are nailed together, they will form the desired inside angle of 135°.

Design P 1490

576 Sq. Ft.—First Floor/362 Sq. Ft.—Second Floor/6,782 Cu. Ft.

Wherever situated—in the northern woods, or on the southern coast, these enchanting A-frames will function as perfect retreats. Whether called upon to serve as hunting lodges, ski lodges, or summer havens, they will perform admirably. The size of the first floor of each design is identical. However, the layouts are quite different. Which do you prefer?

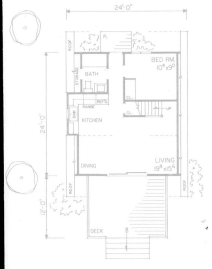

Design P 1491

576 Sq. Ft.—First Floor/234 Sq. Ft.—Second Floor/6,757 Cu. Ft.

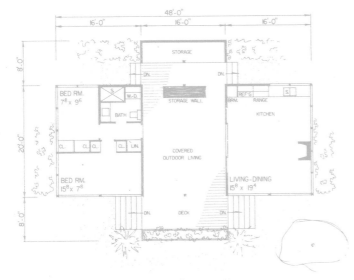

Design P 1417

976 Sq. Ft./8,618 Cu. Ft.

For that forward look the exciting folded-plate roof of this house rests on sturdy posts and beams and extends over a big outdoor living deck that connects a living room-kitchen with two bedrooms and a bath. The deck affords access to the house from both the front and rear, and has two storage walls. Both enclosed areas have large glass walls. Come rain, or come shine, this is a second home you'll have fun living in. It is also one that will be long remembered.

Design P 1401

850 Sq. Ft.—Upper Level
374 Sq. Ft.—Lower Level/10,137 Cu. Ft.

A delightfully contemporary flat-roofed vacation home. The large expanses of glass areas of the upper level allow for an unrestricted view of the lake shore. Study carefully the floor plans at left.

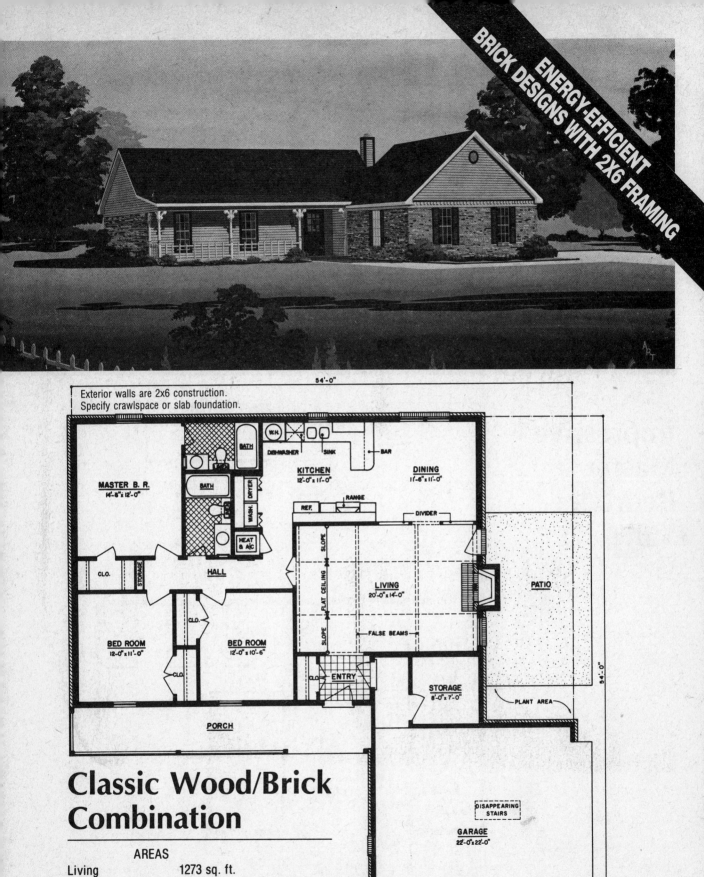

ENERGY-EFFICIENT
BRICK DESIGNS WITH 2X6 FRAMING

Exterior walls are 2x6 construction.
Specify crawlspace or slab foundation.

54'-0"

54'-0"

MASTER B. R.
14'-8" x 12'-0"

BATH

BATH

W.H.

DISHWASHER

SINK

BAR

KITCHEN
12'-0" x 11'-0"

DINING
17'-6" x 11'-0"

CLO.

STORAGE

DRYER

WASH.

HEAT & A/C

HALL

REF.

RANGE

DIVIDER

SLOPE

FLAT CEILING

SLOPE

LIVING
20'-0" x 14'-0"

PATIO

FALSE BEAMS

BED ROOM
12'-0" x 11'-0"

CLO.

CLO.

BED ROOM
12'-0" x 10'-6"

CLO.

ENTRY

STORAGE
8'-0" x 7'-0"

PLANT AREA

PORCH

DISAPPEARING STAIRS

GARAGE
22'-0" x 22'-0"

Classic Wood/Brick Combination

AREAS

Living	1273 sq. ft.
Garage	510 sq. ft.
Storage	56 sq. ft.
Porch	162 sq. ft.
Total	2001 sq. ft.

An Energy Efficient Home
Blueprint Price Code A

Plan E-1212

TO ORDER THIS BLUEPRINT,
CALL TOLL-FREE 1-800-547-5570
(Prices and details on pp. 12-15.)

HomeStyles
SOURCE 1
DESIGNERS NETWORK

Impressive Master Bedroom Suite

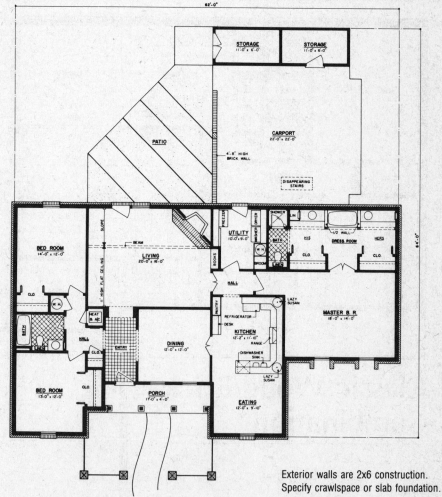

Exterior walls are 2x6 construction.
Specify crawlspace or slab foundation.

Total living area: 1,868 sq. ft.

An Energy Efficient Home

Blueprint Price Code B

Plan E-1818

**TO ORDER THIS BLUEPRINT,
CALL TOLL-FREE 1-800-547-5570**
(Prices and details on pp. 12-15.)

BEST-SELLING
ENERGY-EFFICIENT TWO-STORIES

FRONT VIEW

Week-End Retreat

For those whose goal is a small, affordable retreat at the shore or in the mountains, this plan may be the answer. Although it measures less than 400 sq. ft. of living space on the main floor, it lacks nothing in comfort and convenience. A sizeable living room boasts a masonry hearth on which to mount your choice of wood stove or pre-fab fireplace. There is plenty of room for furniture, including a dining table.

The galley type kitchen is a small marvel of compact convenience and utility, even boasting a dishwasher and space for a stackable washer and dryer. The wide open nature of the first floor guarantees that even the person working in the kitchen area will still be included in the party. On the floor plan, a dashed line across the living room indicates the limits of the balcony bedroom above. In front of this line, the A-frame shape of the living room soars from floor boards to the ridge beam high above. Clerestory windows lend a further note of spaciousness and unity with nature's outdoors. A huge planked deck adds to the indoor-outdoor relationship.

A modest sized bedroom on the second floor is approached by a standard stairway, not an awkward ladder or heavy pull-down stairway as is often the case in small A-frames. The view over the balcony rail to the living room below adds a note of distinction. The unique framing pattern allows a window at either end of the bedroom, improving both outlook and ventilation.

A compact bathroom serves both levels and enjoys natural daylight through a skylight window.

First floor:	391 sq. ft.
Upper level:	144 sq. ft.
Total living area:	535 sq. ft.

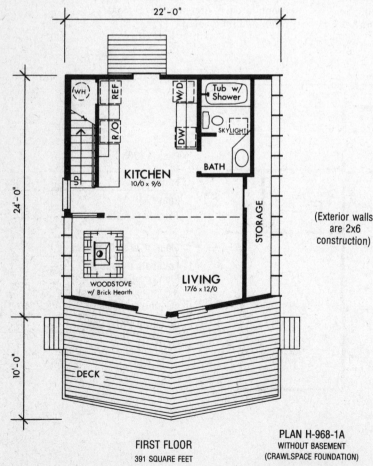

(Exterior walls are 2x6 construction)

FIRST FLOOR
391 SQUARE FEET

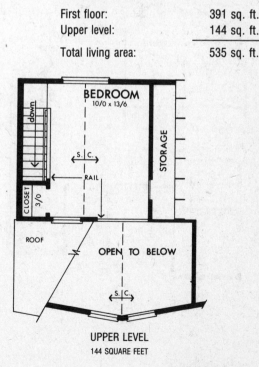

UPPER LEVEL
144 SQUARE FEET

PLAN H-968-1A
WITHOUT BASEMENT
(CRAWLSPACE FOUNDATION)

Blueprint Price Code A

Plan H-968-1A

TO ORDER THIS BLUEPRINT,
CALL TOLL-FREE 1-800-547-5570
150 (Prices and details on pp. 12-15.)

HomeStyles SOURCE 1 DESIGNERS NETWORK

For Luxurious Lifestyles

- Luxurious lifestyles demand more from a new home than space alone. Luxury buyers want volumes with character, impact views, special details, and a well-coordinated design theme, inside and out.
- This plan features an important exterior entry leading into a skylit atrium hall which creates the central formal axis of the home.

- To the left of the atrium is the sleeping zone, including a posh master suite with vaulted ceiling, fireplace, walk-in closet, private den access, and splashy master bath with a tub under oversized corner windows.
- To the right of the atrium is the informal living zone with family room, kitchen, greenhouse breakfast room and mud room.

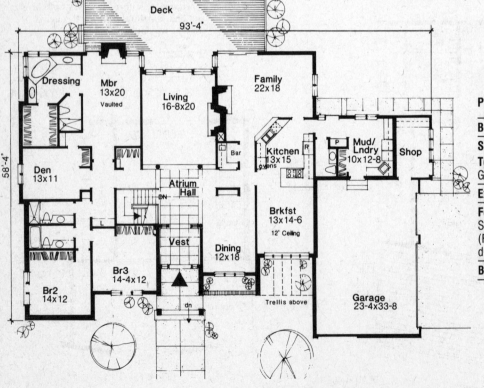

Plan B-1903-L

Bedrooms: 3-4	Baths: 3½
Space:	
Total living area:	3,576 sq. ft.
Garage:	786 sq. ft.
Exterior Wall Framing:	2x6

Foundation options:
Standard basement.
(Foundation & framing conversion diagram available — see order form.)

Blueprint Price Code:	F

HomeStyles SOURCE 1 DESIGNERS' NETWORK

Plan B-1903-L

Rick submitted four sets of drawings: "a structural view based on the span sheet the county gave me; a foundation plan; a floorplan that included electrical and plumbing; and side views that showed the structure in finished form." The plans were approved. "It was the first time I drew house plans, and I was surprised how little there was to it."

For a standard slab foundation, Rick built octagonal foundation forms. In the center he dug out a four-foot square, thirty inches deep, where the weight of the fireplace and beam work demanded a foundation of thick reinforced concrete. During the pour, Rick ringed the perimeter with anchor bolts that would later be used to fasten the walls to the foundation. In the

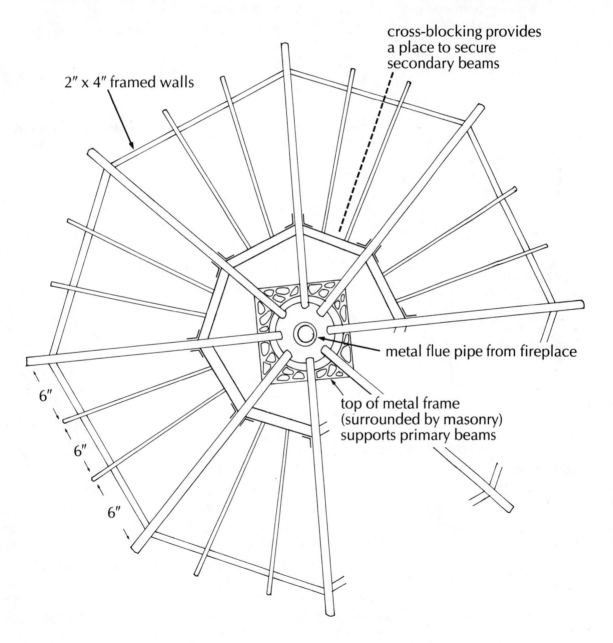

cross-blocking provides a place to secure secondary beams

2" x 4" framed walls

metal flue pipe from fireplace

top of metal frame (surrounded by masonry) supports primary beams

6"

6"

6"

Aerial View of Roof Framing in Main House

49

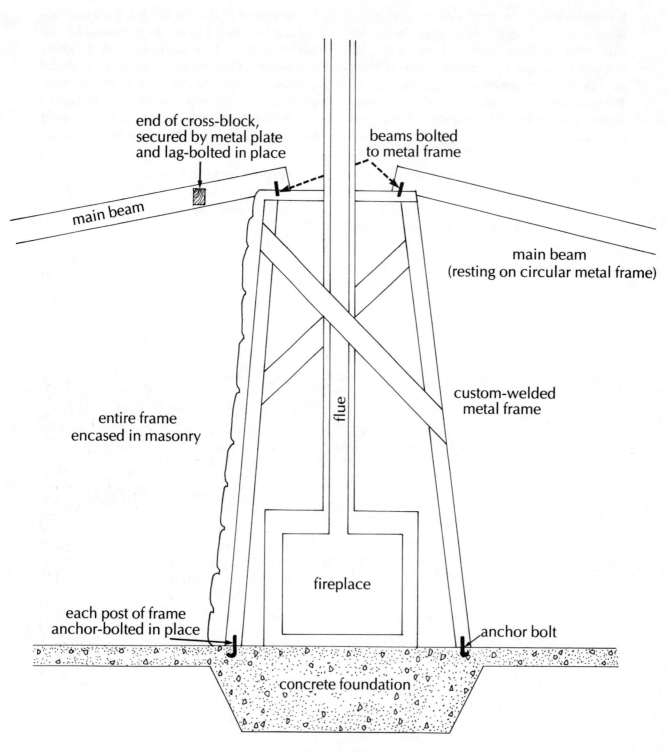

end of cross-block,
secured by metal plate
and lag-bolted in place

beams bolted
to metal frame

main beam

main beam
(resting on circular metal frame)

entire frame
encased in masonry

custom-welded
metal frame

flue

fireplace

each post of frame
anchor-bolted in place

anchor bolt

concrete foundation

Schematic Side View

reinforced center he carefully set more anchor bolts to hold in place a custom-made steel support frame that would serve as the hub for the roofbeams and contain the fireplace.

Rick framed the outside walls with standard 2″ x 4″s and sheathed them with ⅜″ plywood. In the center of the slab he then bolted the 14′ steel frame. Reinforced with welded cross bracing and resembling a small oil derrick, it consisted of four 5″ steel posts four feet apart at the base, leaning in towards the center and united at the top with a steel ring, 3′ in diameter and ¾″thick. This ring is the resting place for the roof beams. Using a block and tackle, Rick bolted eight 6″ x 14″ roofbeams to the top of the ring and to each corner of the octagon. Rick's beams end beyond the junctions of each 18-foot wall; from above, the roof framing looks like the spokes of a wagon wheel.

As Rick points out, "There is a problem in decking over beams that radiate out from the hub. The farther the beams extend from the center, the farther apart they get." He solved the problem by bolting blocking between the beams near the hub. From each block, two additional beams were run to the outer walls. In this manner, he added sixteen more beams to the roof system. "But even with the secondary beams, the decking near the walls would have to span six feet. Two-inch pine will span only four feet, so I used fir, which is strong enough to span six feet."

On the slab inside the center metal frame, Rick placed an airtight wood-burning fireplace. An eight-inch double-walled flue pipe runs up the center of the frame and out the roof. To lay the stone and brick that would encase the fireplace and frame, Rick decided to hire professionals. "The masonry isn't just a fireplace and chimney. It's also a 20-foot partition—the most

Another view of the house, with the hill rising behind

dominant wall in the house. Some people never realize that except for the bathroom, this house is one big room. The partition blocks the view so no more than 25% of the house can be seen at one time. The partition and floor plan make this one-room house as comfortable as most multi-room dwellings of the same size. The entryway, dining area, and kitchen harbor the most disruptive activities—but those areas are all on one side of the stone wall. On the other side of the partition is the living area, and the bedroom and bath are on a raised area behind it."

Then Rick got to work making the house watertight. He covered the roof's fir decking with 1″ rigid insulation and tan asphalt shingles. He installed wood-framed windows, covered the plywood walls with building paper, and applied finished veneers of adobe brick and tan-painted stucco. "I used earth tones because

Freestanding metal stove

the house sits on top of the knoll and I didn't want to offend anyone by forcing them to look at a garish horror. Also, the roof line repeats the slope of the knoll, helping to keep the house inconspicuous."

Inside, the floors are covered in handmade Mexican Tecate tile and oak. Rick wired, insulated, plumbed, sheetrocked, and hired a pro to tape and texture the sheetrock. For trim, he used oak and redwood. Around the chimney, he framed in skylights that allow light to pour in at different angles throughout the day.

Meanwhile, Larry was thinking about creating his space differently: "A friend had stockpiled materials over a period of time and then designed a pretty decent house around those materials. I wasn't in any hurry, so I watched the newspapers and relied on my contacts in the building trades." He concentrated on heavy beams, framing material, wood-framed windows and doors. "I saw a lot of junk, but every once in a while I came across some good-quality material at good prices."

While Larry was collecting materials, he built a 10′ x 16′ houseboat to live on. During the winter, the structure—made of plywood and discarded windows, lashed to ten 50-gallon drums—floated in the tules and bullrushes at

Larry's salvaged building materials, before construction began

the sunny end of the pond. In early spring Larry moved it under a shady oak tree at the pond's other side. Though Larry's recollection of the houseboat is favorable, he found the first spring and summer a little unpleasant: "The bugs were nasty. Friends stayed away. I imported a lot of frogs and small-mouth bass and channeled some spring water into the pond. By keeping the pond high enough to support aquatic life, there were fewer bugs in the following years." At the end of summer, the pond always shrank to the size of a large puddle, and the houseboat sat on parched, cracked mud.

After three years of houseboat living, Kathy moved in—spurring Larry into action. "The houseboat was too cramped for two. Kathy's arrival pushed me into taking stock of my materials and thinking about a 'real' house. I'd amassed enough stuff to get started. I had most of the framing, all of the windows and doors, and Spanish tile roofing. I wanted to make a straightforward, energy-efficient house."

Site selection was important: "A lot of people buy and build simultaneously. They may luck out, but more often they don't use the best combination of the sun's influence and their topography. I needed to make good use of the winter sun, and living on the land for three years helped me pick the best spot."

Larry rented a light tractor and scraped a flat area for a pad slightly into a south-facing slope. He planned to nestle his north wall into the slope to reduce heat loss. Since the site was hidden from other homes and was inaccessible by car, Larry decided to build without a permit. "I adopted a 'build first, worry later' philosophy but planned to follow the codes in case my house should come to the officials' attention." Since the site was also inaccessible by cement truck, "I rented a small mixer from an equipment-rental firm. With the help of friends, we were able to pour and smooth the entire slab in about ten hours." The foundation was built to code with a plastic water barrier and rebar.

Larry's framing plan saved a lot of money. "Because I had collected so many large timbers,

Recycled wall panel

post-and-beam seemed the way to go." Larry placed his posts to accommodate the dimensions of windows and doors, filling in most of the spaces in between with homemade adobe brick. "I experimented a little to get a brick that wouldn't dissolve in a bucket of water, and could also be cut with a graphite blade in a skillsaw." It took about a week to make enough bricks with his final formula of 30% clay, 60% sand, and 10% Portland cement.

In a corner of the main room, Larry installed a kitchen with cabinets he got from a remodeling job. Behind the main room, he built a 10' x 11' bedroom and a walk-in closet made from a finely finished 8' x 10' cookie wagon that had once been pulled by horses in Disneyland. You walk through the "closet," with its contoured birch ceiling and hardwood floor, to get to a full bathroom.

Larry installed a minimal septic system. The leach field he dug by hand along with the hole for a 750-gallon, cinder-block septic tank. The toilet empties into the septic tank; all the other drains channel through a gray-water filter to a garden. Larry enjoyed plumbing: "It's like working with an erector set. Even copper pipe is fun; once you learn to solder, it's a cinch." Larry buried the incoming pipes along the side of the house and brought them through the walls where needed.

To supply hot water, he installed a solar

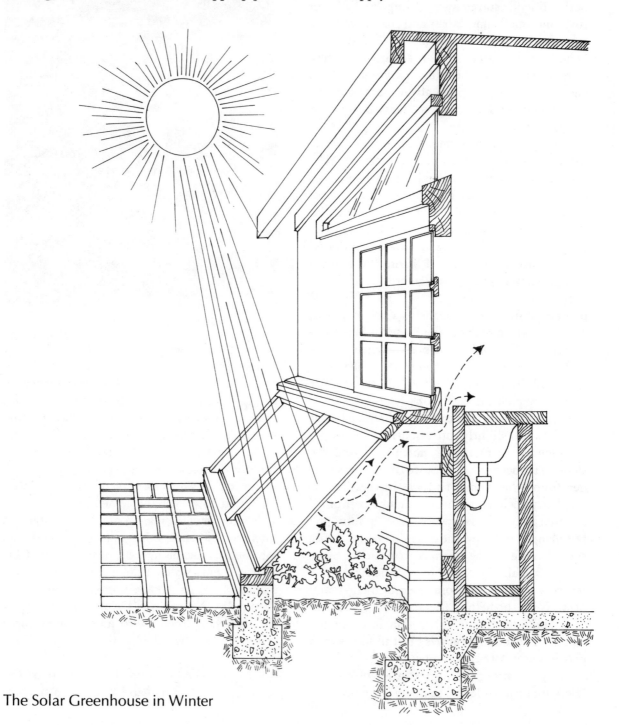

The Solar Greenhouse in Winter

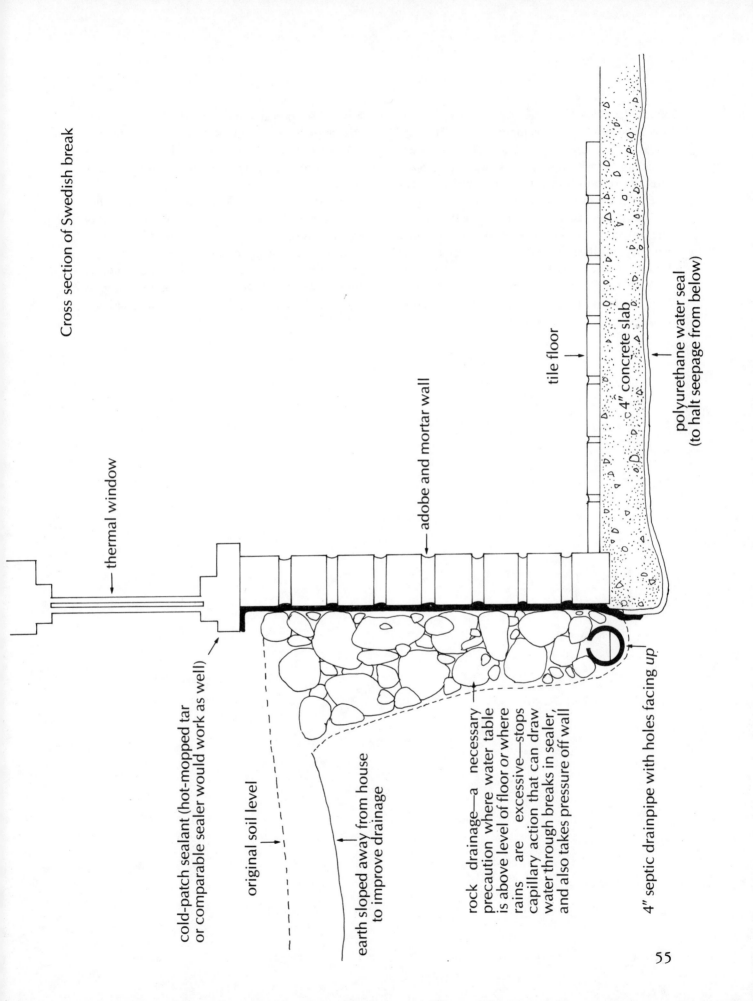

Cross section of Swedish break

thermal window

adobe and mortar wall

tile floor

4" concrete slab

polyurethane water seal
(to halt seepage from below)

cold-patch sealant (hot-mopped tar
or comparable sealer would work as well)

original soil level

earth sloped away from house
to improve drainage

rock drainage—a necessary
precaution where water table
is above level of floor *or where
rains are excessive*—stops
capillary action that can draw
water through breaks in sealer,
and also takes pressure off wall

4" septic drainpipe with holes facing *up*

55

on the door. Kathy: "The inspector was lost! He was looking for a house under construction and found us by mistake. He asked if we had a building permit, and of course we didn't. So he cited us for building without a permit. A week later, the county told us we'd have to tear down the house. Since Larry was legally half-owner of the property and Rick's house was a sanctioned primary dwelling, we took the position that our house was the guest house.

"Luckily, our house is small enough to qualify. The county haggled with us, but they eventually gave in. Larry was fined double the cost of a building permit, and they made us take out the range because a guest house can't have one. We were allowed to continue living here because a property owner can live in his own guest house."

Larry feels the whole episode was somewhat of a fluke, but he's glad he thought out his strategy in advance. "If the house hadn't been built to code and if we hadn't been able to get it classified as a guest house, they could've made us pretty miserable. It shows if you're going to build without a permit, it's best to have a plausible defense."

10. A Passive Solar Home and Nature Conservancy

RHODES SCHOLAR, playwright, college professor—Jim Bierman is a man of high ideals and accomplishments. On vast acreage overlooking the Pacific Ocean, he has realized his dream of a self-sufficient passive solar house.

In the early 1970s, Jim was hired as an assistant professor at a coastal university. In his free time, he went hiking in the nearby mountains.

"Regardless of where I started or planned to go," Jim explains, "I always ended up on the same knoll where hawks soar; where spectacular, rolling, grassy, partially-forested hills descend to a white ribbon of beach and the Pacific Ocean. One day I noticed a 'For Sale' sign. I phoned the realtor and found my favorite spot was part of 326 acres being sold for $320,000."

The passive solar home—side view

breadbox heater. When it came to warming the air, Larry came up with a number of innovative ideas. He installed a freestanding metal fireplace and ran the flue pipe only 8″ from a solid adobe wall. "The wall collects heat from the pipe and becomes a heat source itself." Outside, below the kitchen window, Larry designed a mini-greenhouse with removable panes. During the winter he leaves the panes in place; warm air accumulates and enters the house through a vent above the kitchen sink. In summer, the panes are stored and the vent closed.

On the outside of the north wall, Larry plastered over the adobe brick and painted the bottom three feet with cold patch (an oil-based roof-patching material). Larry had seen some signs of ground water in the slope when he'd excavated the site. "I wanted to backfill against the north wall, but I was hesitant because capillary action in ground water can pull water through the smallest split in the waterproofing sealant. Before I backfilled, I decided to build a Swedish break against the wall."

Swedish breaks dissipate capillary action and divert ground water away from a wall, and there are different ways to build them. Larry placed a 4″ perforated leach pipe, holes up, along the base of the wall. "Then I shoveled about 30″ of drain rock onto the pipe and against the wall. I backfilled the natural soil against the rock and put about 4″ over the top of the rock. Any water coming through the ground trickles down through the rock and flows out the pipe and down the bank." Larry made sure the surface water would flow away from the house by sloping the backfill against the wall. His efforts paid

The greenhouse/heat collector in summer—with its panes removed

Vent from the greenhouse/heat collector

off: moisture has yet to penetrate the north wall.

How Larry got the green Spanish tile roofing is a renegade builder's dream: "My mother met a rancher who had bought a house with a green tile roof. The guy hated the roof so much that he took it off and replaced it with red Spanish tile. I phoned the rancher, and he seemed surprised anyone would want a green roof and told me to come and get it—free. When I arrived, he looked at me and said, 'I don't know what good a green Spanish tile roof is, unless you're color blind!' We both laughed—the rancher because he enjoyed his joke, and me because I couldn't believe what a deal I was getting."

For three years, Larry and Kathy lived comfortably without electricity. Besides the solar water heater, they used a backup propane water heater and a propane range and refrigerator. For lighting, they used kerosene lamps. When Lar-

ry got around to bringing electricity into the house, he ran a direct burial cable from Rick's house.

Larry's house was finished except for the Tecate tile floor. Rather than disturb the completed walls, he decided to work the wiring into the floor, near the walls, where there would be no traffic. After laying the 12″ square, 1″ thick Mexican tiles, he ran Romex wire in the grout line and grouted over it. This isn't the customary or the code way to lay wire, but in his case, Larry feels it's safe. "I only have 110 amp circuits in the house, so under most situations, there isn't enough juice in any one wire to do too much harm. I just made sure the wire sits perfectly flat, and it's well covered with grout."

About a year after Larry's house was totally finished, his "build first, worry later" philosophy met its test. A building inspector knocked

The price was prohibitive, but Jim devised a purchase plan: "Seven of us pooled our resources and were able to buy the property. I was concerned about preserving the environment's integrity and ensuring that each of the buyers had the opportunity to enjoy the entire parcel. Besides abiding by all the stringent governmental guidelines for developing a parcel on the California coastline, we wrote a charter to deal with the property's future. There are five provisions: 1) No house can be built within view of another house, 2) All owners will have recreational use of the entire 326 acres, to within fifty yards of any house, 3) There will be no hunting, firearms, or off-road driving on the property, 4) The entire 326 acres becomes a permanent nature conservancy, and 5) Future owners will honor all aspects of the charter."

As it turned out, the land has some financial benefits. A local farmer pays $5 a head to graze his cattle on the property, and the county designated the wooded areas as tax-exempt timber reserves.

Naturally, Jim's personal forty acres included his favorite knoll. He decided to build an inconspicuous, partially subterranean, passive solar home. But first he spent a year researching and consulting experts. "I learned to be wary. Solar construction is a new field, and there are charlatans who claim to be experts. I chose a contractor who had published several articles on the subject, but unfortunately, he was long on theory and short on practical knowledge. He screwed up badly, and we had to tear out most of the first foundation because it wasn't strong enough."

Once Jim found someone with the expertise to do the job, it took about four months to complete a 1,300-square-foot masterpiece, with a simple plan: "We built a rectangular box and tucked the long north side into the hillside. The sloped roof is covered with sod. The south side is mostly insulated glass that lets in the sun's energy, but is reluctant to let it out. The bottom floor is concrete, covered with brick that collects and saves the sun's energy to release at night."

For the greatest structural strength, Jim had the slab foundation and the north wall poured as a single unit. The north wall is reinforced concrete because it has to serve as a retaining wall against tons of backfill. By itself, a buried concrete wall is not waterproof, so before backfilling, Jim had roofers "hot mop"—apply hot tar to—the outside of the wall.

Besides being an excellent insulator, the sod roof blends into the landscape so well that Jim is used to guests asking him where the house is. From the north side, all you see are a few vent pipes and a chimney poking through what looks like a rectangular knoll. Sometimes the roof fits the natural scheme *too* well: "One afternoon, I found a cow grazing in the middle of my roof, so I fenced the cattle away from the house. Deer still wander onto the roof, though."

Sod may fit into the aesthetic, energy-efficient design, but it was not a low-cost item. "You need heavy laminated beams to hold up sod a foot thick. And the expensive tar roof that has to be installed under the sod pushes costs up about $5,000 over a conventional well-insulated roof."

On the plans submitted to the County, Jim designated the second of the house's two stories as storage space. This way, he was able to install

Bread oven in flue provides additional heat

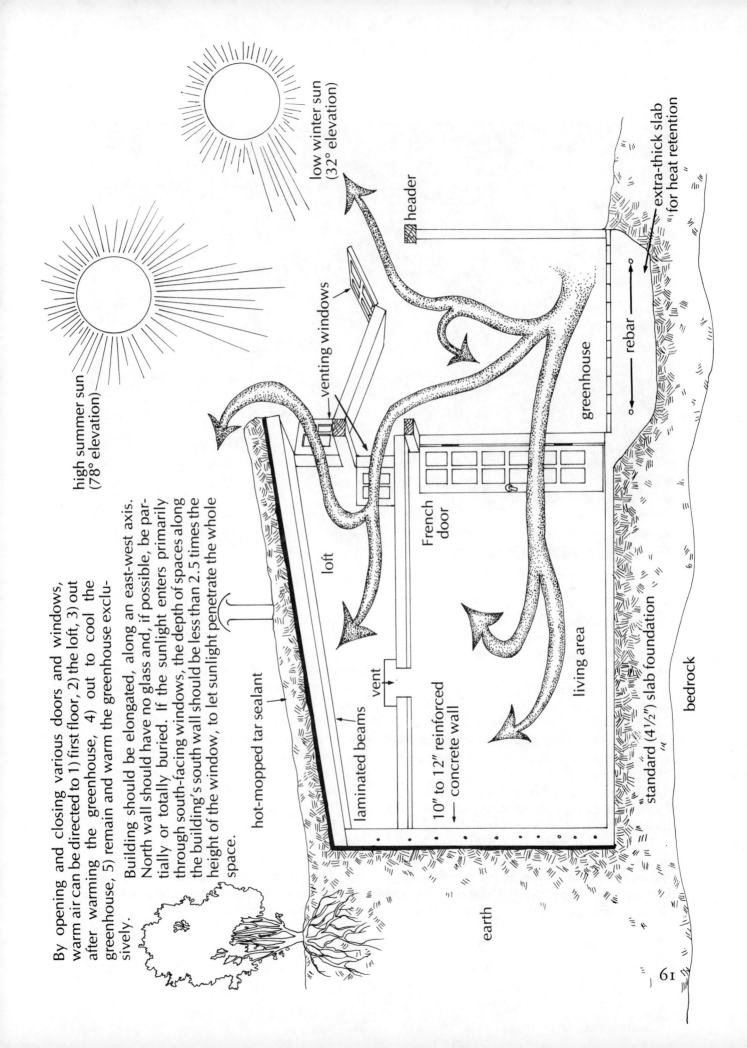

By opening and closing various doors and windows, warm air can be directed to 1) first floor, 2) the loft, 3) out after warming the greenhouse, 4) out to cool the greenhouse, 5) remain and warm the greenhouse exclusively.

Building should be elongated, along an east-west axis. North wall should have no glass and, if possible, be partially or totally buried. If the sunlight enters primarily through south-facing windows, the depth of spaces along the building's south wall should be less than 2.5 times the height of the window, to let sunlight penetrate the whole space.

high summer sun
(78° elevation)

low winter sun
(32° elevation)

header

venting windows

loft

French door

vent

laminated beams

hot-mopped tar sealant

10" to 12" reinforced concrete wall

living area

greenhouse

rebar

extra-thick slab "for heat retention

standard (4½") slab foundation

earth

bedrock

Vent from living room to second-floor loft

a circular staircase four feet in diameter instead of the more expensive eight-foot model. The spacious upstairs can be used for a study and bedroom.

With self-sufficiency in mind, Jim designed an electrical system that can be powered by either conventional power or a 12-volt battery. "Living on the coast guarantees a steady breeze needed to run the wind generator that keeps the batteries charged. Because a battery system can't lose more than three percent of its voltage, I used 6-gauge copper wire throughout most of the house, instead of the standard 12-gauge."

The washing machine, refrigerator, water heater, and range all run on propane. Jim found the antique gas range rusting in a friend's backyard and fixed it up for $150. Other friends gave him the refrigerator. For about four months of the year, the water heater runs on propane; the rest of the time it is solar-powered.

On harsh winter nights, Jim relies on two forms of wood heat: "I built the fireplace with stones from the creek. Actually the fireplace is more decorative than efficient, because open flues suck more warm air out of a room than they put into it. So I installed an intake vent that allows the fireplace to draw air from the outside, which helps somewhat." For really efficient use of wood, Jim installed a small free-standing metal stove with a bread-baking oven in the flue. For maximum heating, he opens the oven doors.

Throughout the house, Jim has taken great care with the finishing touches. The kitchen countertop was specially milled from a single redwood slab. The bathroom features an an-

Upstairs sleeping loft

Stone fireplace: vent on hearth sucks in air from outside

Solar water-heating panels and outdoor shower

tique toilet, complete with head-high water closet and polished brass pipe. The indoor shower is decorated with tile; outdoors is a solar-heated shower.

Like most owner-builders, Jim took an active role in the creation of his house, doing most of the planning and much of the work. He conscripted friends and builders who wanted to learn more about solar construction. But Jim feels that amateur builders should stay away from certain facets of the job: "It's important to recognize when to call in an expert, because in some areas, getting it 'almost right' isn't good enough. For the foundation, north wall, and roof, I felt it was important to get the best people I could find. But sometimes I overstepped the bounds, because I wanted so badly to be part of the process."

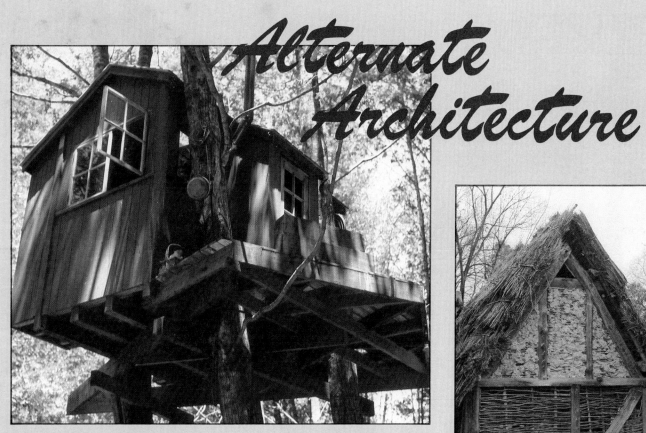

Alternate Architecture

A Walking Tour

Clockwise from top left: *a treehouse with running water; daub-and-wattle construction; wooden yurt; and a renovated lighthouse*

ALL PHOTOS: ERIC HOFFMAN

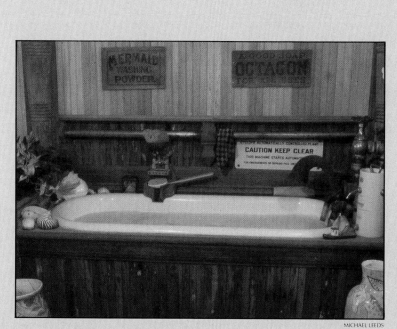

MICHAEL LEEDS

Above: *second-floor bathroom, with enclosed free-standing tub, of the converted
19th-century warehouse described in Chapter 5*

BOB BARBOUR

Above: *exterior view of the structure, with its
original wooden facade*

MICHAEL LEEDS

Above: *close-up of the pressed tin ceiling and
cornice during installation*

Below: *the late Victorian double bed*

MICHAEL LEEDS

Right: *pagoda of bare beams shelters a patio/entryway*

Below: *the stamped and dyed concrete floor discussed in Chapter 20*

MAGNUS BERGLUND

MAGNUS BERGLUND

MAGNUS BERGLUND

Right: *installing the sod roof*

Below: *exterior view of a rammed-earth house, showing the greenhouse and spa*

MAGNUS BERGLUND

Above: *bending the structure's rebar skeleton*

Above: *the "periscope" with its metal mesh and rebar in place*

Right: *the freshly-"mudded" skeleton (note braces to support wet cement and guard against sagging)*

ALL PHOTOS BY CECILE CHAMPAGNE

Left: *the original clay model of the structure*

Clockwise from top left: *circular staircase in the ferrocement studio described in Chapter 18; French doors with ferrocement "awning"; the downstairs, as seen from the bedroom; one of the periscopes that serve as vertical skylights; exterior view*

MARY GORDON

BOB BARBOUR

Left: interior of the passive solar home described in Chapter 10

Below left: another view of the dining nook

Below: the exterior, showing the second-story deck and sod roof

BOB BARBOUR

BOB BARBOUR

Below: the south-facing exterior, as seen from the ground

BOB BARBOUR

CECILE CHAMPAGNE

Left: *stained-glass windows and free-form bed belie the fact that this bedroom is actually afloat—in a houseboat moored at Sausalito*

CECILE CHAMPAGNE

Right: *a "subdivision" of distinctly different houseboats resting at permanent anchor*

CECILE CHAMPAGNE

Left: *the kitchen, fitted against one of the houseboat's exterior walls*
Below: *the houseboat's bathroom, fitted into yet another corner*

CECILE CHAMPAGNE

CECILE CHAMPAGNE

Above: *the original water tank (discussed in Chapter 3) converted into an art studio*

CECILE CHAMPAGNE

Above: *the converted water tower with its "apron" of rooms, as seen at a distance*

Below: *the front door, sheltered by vines and other greenery*

CECILE CHAMPAGNE

AUTHOR PHOTO

Above: *skylight in the ceiling of the water tower/art studio*

11. Building
for a Sense of Community

AFTER GREG and Chris graduated from college in 1977, they went different ways. Chris became a designer and building contractor and Greg a radio producer. But after four years of work in their respective fields, neither could afford to buy a house. Chris: "I had moved into a large apartment complex and couldn't stand the sterility and emptiness of the place.

Three years after graduation, I kept remembering the feeling of community I had in college, when I lived with people who were alive and had similar points of view. I remembered how a group could rent a much nicer house than an individual could. So I decided to combine the positive aspects of my college experience with permanent home ownership. I contacted Greg,

One of the "community" cottages

Cottage with glass overhang and sliding insulated shutters

and he felt the same way. He had rented a house with a couple of friends, spent time and money fixing it up—only to have the landlord sell it. We had different experiences, but they added up to the same miserable thing. It didn't take us long to find eight other people who shared the housing blues."

Greg and Chris organized a meeting to explore group home ownership. All ten people discussed priorities and found they wanted 1) a sense of community, 2) attractive housing, 3) a beautiful site, 4) privacy from one another, 5) commitment to energy efficiency, and 6) an organization to deal with ongoing financial matters and questions affecting the undertaking.

After a couple of months of scouring the housing market, they found plenty of two-bath, three-bedroom houses—mostly at 17% interest in the $150,000 to $200,000 range. Greg: "We knew we could do better if we built something ourselves, especially since Chris is a contractor and Bob a carpenter."

Chris and Greg called another meeting and found that the entire group would prefer to build rather than buy something that fell short of their ideal. They agreed that their venture should include a small cottage for each individual and one large central structure with a kitchen, recreation facilities, and library/media center.

They began looking for property. Chris found a ramshackle two-bedroom house on ten acres, with a lovely view, priced at $135,000. Seven of the group's ten people wanted to buy the property. Another meeting was held to draw up formal rules for buyers. They agreed that 1) there would be seven equal shareholders, 2) a shareholder could have a spouse or mate as a

74

partner, to share whatever the shareholder was entitled to, 3) only shareholders could vote in meetings concerning the venture, though spouses' or mates' views were welcome, 4) if a shareholder wanted to sell, the group had the first right of refusal, and the seller was responsible for finding an acceptable buyer, 5) the seller could not make more than 10 percent profit in a three-year period, and 6) shareholders doing more than their share of construction would be granted credits to compensate for extra labor.

Each buyer put $5,000 towards the down payment, then negotiated individual personal loans for the balance of the purchase price. Their monthly payments ranged from $150 to $300. Greg: "Everyone elected to live in the existing house rather than waste money that could be used for construction." So Greg, Chris, Mark, Charles, Nikki, Bob, and Karen moved into the leaky, termite-infested, two-bedroom dwelling. These cramped conditions spurred construction of the structures that would make up the community. Mark: "Among the seven of us, we had a substantial income. We planned to finance most of the construction costs out-of-pocket. As it turned out, we also used family loans, and Chris's ability to locate and bid on used materials helped finish the project successfully."

To address expenditures and other matters, the group held regular meetings. In the first one, Chris and Bob presented a number of building designs and costs. "We elected to build seven cottages, each with about 530 square feet of floor space. This size was about

The kitchen, with communal wok

The main house, showing plenum and duct system

right because we could get the finished shell up for about $5,000, which was an acceptable price for everyone."

About the time the group was ready to start construction, voters passed an ordinance limiting growth throughout the county. Anyone wanting to build had to put up $500 to put their name in a lottery and hope to be awarded one of the 500 annual housing-starts. If you weren't chosen, you could always try again next year. In addition, a state referendum put stringent restrictions on developments within view of the California coast. Even if they won in the lottery, they would still face the Coastal Commission—another six month's delay. Greg: "All the red tape was a real blow. We were already living in crowded conditions, and the bureau-

cratic regulations screwed up our financial timetable and jeopardized the entire project. We discussed it, and decided to go ahead without permits. Building without a permit isn't treated too harshly as long as you conform to building codes and zoning ordinances. So we decided to stay entirely within the building codes, documenting each step with photographs."

Improbably, their plan fit the ordinances, since their property was zoned for a single-family residence and secondary structures. The only kitchen—the distinguishing feature of a single family residence—would be in the central structure, which meant all the cottage/residences could be viewed as secondary structures. Greg: "Our choice of a barn motif also diminished the chances of a confrontation with county officials. Traditionally, farmers in the area have built barns and other extra buildings with impunity. Most building departments lack the manpower to go out looking for code violations; they just react to complaints turned into the department. So to be on the safe side, we discussed our plans with our two closest neighbors."

Once actual construction got underway, everything went smoothly. "Chris and Bob organized our work effort. We all pitched in."

First the group put down a slab foundation and framed the structure with 2″ x 6″ material. They opted for 2″ x 6″ instead of the usual 2″ x 4″ because a thicker wall could contain more insulation. The exposed ceilings were supported on simple trusses built by Bob and Chris. They laid an inch and a half of rigid insulation over the roof decking and roofed the building with sheets of heavy-gauge Curoca roofing. After the first cottage was up, the rest were fairly easy because they're all similar. Each shell took about five weeks to construct.

Everyone pitched in to cover most of the walls with sheetrock that is taped and painted white. Recycled hardwood or one-foot-square tiles were laid on the floors.

Though the cottages are similar in design, there are major differences. Some are two-story, others one. Chris purchased all the doors in a carriage-house auction, but no two cottages have windows of the same size. Most are small-paned, and many of the wooden window frames

Catwalk leading through trusses

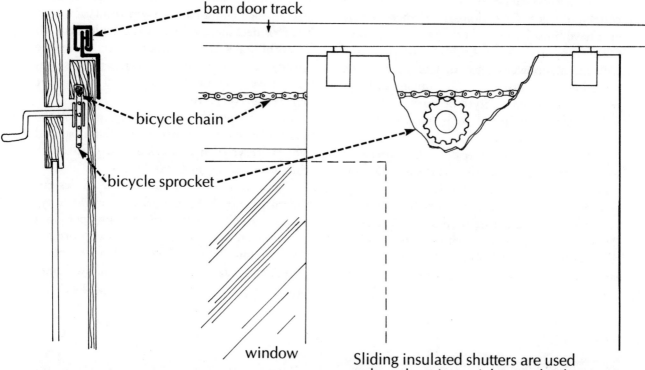

barn door track

bicycle chain

bicycle sprocket

window

Sliding insulated shutters are used
to keep heat in at night—and to keep out
excess sunshine during the day.

were made on the property. Bob: "Chris and I collected gray shorts and sticker wood from local lumber companies who sold it as firewood. With our planer, we made perfectly adequate window frames." One cottage has shutters that are cranked manually along a garage-door track. Chris happened upon a large greenhouse that was going to be bulldozed and took it apart in sections. Now all the cottages have glass-covered patio greenhouses; two are joined together by a greenhouse in the middle. Mark partitioned his cottage, making two, small, easy-to-heat rooms and an indoor garden that provides year-round vegetables.

While building the cottages, Chris kept an eye out for used materials that could be used in the different structures. He bought part of a fire

escape from a 1930s skyscraper and modified it into a stairway to his loft: "The nice thing about not building with a permit is you have more flexibility. You aren't tied to a set of plans and specific materials. If you come across outstanding, inexpensive used material, you can rethink your plans without reentering the red-tape maze. For us, the important find was eight fifty-foot trusses. A Greyhound bus garage had burned down, but maybe half the trusses made it through the fire. I offered the salvage company $500, and to my surprise, they accepted. Probably so few people want used trusses of that size, they thought something was better than nothing. For another hundred dollars, they even delivered them."

Chris hired a guy with a 40-ton crane. "He

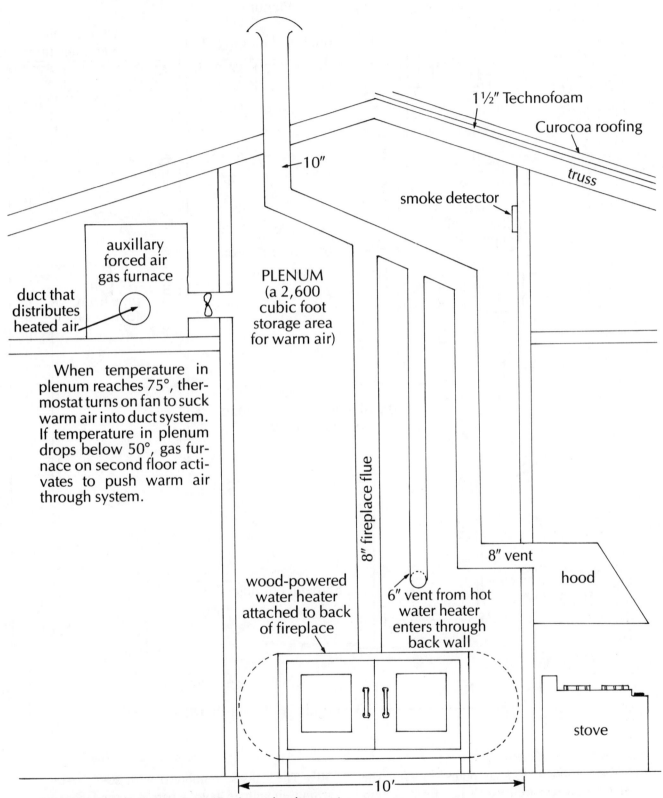

1½" Technofoam

Curocoa roofing

truss

10"

smoke detector

auxillary forced air gas furnace

duct that distributes heated air

PLENUM (a 2,600 cubic foot storage area for warm air)

When temperature in plenum reaches 75°, thermostat turns on fan to suck warm air into duct system. If temperature in plenum drops below 50°, gas furnace on second floor activates to push warm air through system.

8" fireplace flue

wood-powered water heater attached to back of fireplace

6" vent from hot water heater enters through back wall

8" vent

hood

stove

10'

Front of metal fireplace is flush with plenum's exterior walls. Doors open into main room.

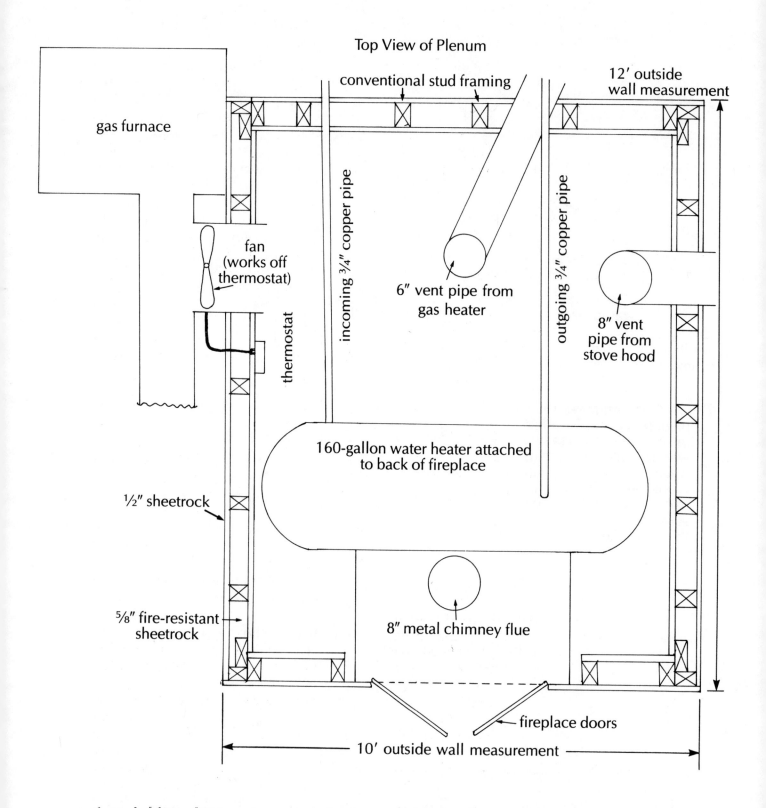

Top View of Plenum

gas furnace

conventional stud framing

12' outside wall measurement

fan (works off thermostat)

thermostat

incoming ¾" copper pipe

6" vent pipe from gas heater

outgoing ¾" copper pipe

8" vent pipe from stove hood

½" sheetrock

160-gallon water heater attached to back of fireplace

⅝" fire-resistant sheetrock

8" metal chimney flue

fireplace doors

10' outside wall measurement

charged $600—$150 an hour—to raise the trusses, and was well worth it." Built around the trusses, the main building ended up 50' x 50'. The trusses, reinforced with steel, were strong enough to hold a partial second floor. Under Chris's guidance, the group poured a

2,500-square-foot slab and framed the walls in either 2″ x 8″ or 8″ x 8″ post-and-beam. "We sheathed the trusses in 2″ x 8″ fir and insulated and roofed the building the same way we did the cottages."

Group workers built an inside stairway to a second-floor catwalk that passes through the center of the trusses and leads to a 600-square-foot room. The kitchen is impressively outfitted with an eight-burner propane range, a huge stainless steel refrigerator, 30 feet of freestanding Formica-covered kitchen cabinets, and, in a separate room, a dishwasher large enough to wash 200 place settings at once. All of this was purchased at ridiculously low cost. Chris: "Restaurants go out of business every day. A friend in the restaurant business told me there are regular auctions; he's seen some appliances move to as many as four restaurants in a five-year period. In hard times, even more restaurants go belly up. We bought most of our kitchen appliances for twenty-five cents on the dollar."

In the central structure, Chris, Mark, and Charles built a multi-phase heating system. In the center of the structure, they framed a 12′ x 10′ shaft—a space containing over 2,600 cubic feet that runs from the floor to the ceiling. Enclosed, the shaft becomes a plenum, an enclosure designed to trap heat and serve as a warm air reservoir. Chris had a custom metal fireplace welded up and inserted into the plenum. The fireplace's decorative glass doors open into the main room, while the rest of the fireplace and the metal chimney are inside the plenum. To the back of the wood-heated fireplace, Chris welded a 160-gallon water tank that works in conjunction with a 200-gallon gas water heater. Before venting through the roof, the wood stove's chimney pipe, the pipes from the gas water heater, and the stove vent all enter the plenum, contributing their heat to the shaft.

At the top of the plenum, where the air is warmest, Chris connected a 14″ duct system. From it, he and Bob installed over 230 feet of duct pipe. When the temperature in the plenum rises above 75° F., thermostat-triggered fans suck the air into the ducts and distribute it throughout the house. If the indoor temperature drops below 50° F., a forced-air furnace connected to the ducts pushes gas-heated air through the system. If the temperature in the house is too high, a manually-operated vent at the top of the plenum can be opened. Greg figures about sixty percent of the heated water and air is generated by wood heat.

Chris bought the ducts and most of the wiring from a company that had been awarded a bid to take down a large building: "A lot of people won't salvage wiring, which surprises me. Wires make the same runs between sockets, regardless of the structure. It's easier to pull out of a house than it is to put in. I got the ducts because the company that was taking the structure down was interested in replacing the building, not the salvage."

About the time the central structure, now dubbed the Main House, was completed, the group tore down the original house and hauled its non-burnable parts to the dump.

In the Main House, the 3,100 square feet of floor space is cleverly put to use. The upstairs is a library and media center. The downstairs has a spacious kitchen, dining area, bathroom, and fireplace seating area. Partially finished are the recreation center (including a gymnasium and Jacuzzi) and production studio. Since most of the residents are involved in television and radio production, they would like to develop their own at-home company.

How does it feel to be part of a small community designed and built by its inhabitants? Greg: "The combination of the cottages and Main House gives us a dimension none of us could have afforded individually. We have developed a community around a theme that makes sense to us. There's sustenance from meaningful friendships. In an inhospitable economy, it's reassuring to be a member of a responsible group that's economically united. Regardless of our individual ups and downs, being tied to a group with common interests guarantees we'll make it."

12. Operating in the Gray Zone

M ANY OWNER-BUILDERS in low-traffic areas debate the pros and cons of obtaining a building permit. A house built with a permit will keep you out of trouble with officials, since an angry neighbor won't be able to report you for violations. A house built without a permit may be difficult to sell—unless you plan to live in it forever. A house built with a permit will qualify for a mortgage—but without a permit, you may be able to build cheaply enough that you don't need a mortgage.

Most builders I talked to either conformed to building codes, or came close enough to make what they had done debatable. Ted—plumber, botanist, volunteer fireman, and leader of most of his rural community's farthest-out events—

Leaded window

offers a unique view: "To me, land ownership is personal freedom with limitations. It will end, just as surely as our lives will end. Our structures should be temporary, like our lives. If all my buildings burned down tomorrow, I'd get to try out some new ideas."

With this introduction, Ted invited me to inspect the structures on his five-acre spread. "Welcome to the land of Mondo Bizarro," he chuckled. "I call most of my structures California Shack Motif."

In 1971, Ted bought fifteen acres for $8,000. In 1976, he sold off ten acres for $25,000. Because of a distinct dislike for red tape and the permanent structures required by building codes, Ted decided against building a "real" house.

So his first home was an 18′ x 22′ equipment shed from an abandoned oil field. "The field foreman was an old friend, so we worked a deal. I bought the shed for $250 and carted it away on a weekend."

The shed, sheathed in the original galvanized tin, came with three sliding windows and a door. Ted reconstructed it with conventional 2″ x 4″ framing on a concrete slab foundation and sheathed the inside with plywood. With the addition of a sink, wood-burning stove, minimal wiring from his well, and a detached outhouse, the shed served as Ted's home for three years.

"The Palace," Ted's present-day house, is perched on a steep slope with a vista of the surrounding hills and coastline twenty miles away. Listed in county records as a storage building, The Palace embodies a philosophy, a personality, irreverence, and good luck. "I really didn't have a plan. It sort of grew out of the materials I was able to get. Over a period of a few years, I collected about twenty telephone poles of varying lengths. I concreted them into the ground in a rectangular pattern with no pole further than ten feet from another. For about a year, I left them there as sort of a testimonial to the telephone pole. Then I did some plumbing for a guy who paid me in girder material.

"I bolted the girders about nine feet up the telephone poles and laid 2″ x 6″ fir across them to make the second-story floor. At ground level, I smoothed the earth around the poles and laid brick that I found at an abandoned kiln."

When Ted set his sights on making a third floor, he found some of his telephone poles were too short to support another level. "So to get the height I needed, I just cut a notch in the short poles and spliced on another section of pole. The third-floor platform was built much like the second floor's."

Now Ted had lots of poles and three floors— but no roof or walls. He took a break: "I went to Europe for a year. Sitting out over the winter gave the floors a little more character."

Upon his return, Ted pushed for completion, starting with the roof. "I traded a guy some plumbing for water-damaged pine decking and rafter material." The rafter system, best described as haphazard, is a matrix of telephone poles, railroad ties and 2″ x 4″s that makes a mockery of architectural design. Ted admits his roof support system—two ridge beams, two feet apart, one slightly higher than the other— is a little different, "but it doesn't move *much* when you walk on it, and the roofing material helps hold it together." Decorating the roofline is a ball from a Japanese fisherman's net. Ted has transformed it into a huge eye, complete with sheet metal eyelashes.

For the roof, Ted used 3′ x 10′ sheets of eighteen-gauge galvanized metal. "Even though it's not really roofing material, I couldn't pass it up at eight bucks a sheet. A sheet metal worker pressed bends into the sheets so I could lock them together. It cost about $160 to cover 600 square feet."

Ted still needed walls. "I put a lot of windows on the south and west sides. Most of them were traded for or salvaged, but my best deal was a 4′ x 10′ leaded window I got for $16 from a contractor who went bankrupt. Around the windows I stuccoed over chicken wire tacked to plywood.

"On the north side I wanted good insulation. The areas between the poles I filled with sheets

Two telephone poles spliced together

"The Palace"

of rigid foam insulation; then I stretched tightly-strung mesh over both sides of the insulation, fastening it to the poles." Using a concentrated mortar mix that "looked right," he plastered the mesh—an idea he got from helping a friend work on a ferrocement boat. The result: a wall that is thin, well-insulated, and extremely strong.

Ted's favorite wall has opalescent glass paperweights embedded in it. "I was given all this beautiful glass, and I had to use it somewhere. I set a rectangular wooden frame on a bed of sand.

To the frame, I attached vertical and horizontal rows of wire, making three-inch squares. In each square, I put a paperweight, then I poured in grout. The wire united the frame to the slab and added strength to the mortar. When it dried, I framed the slab into the wall."

About the time Ted enclosed the Palace, signs of civilization began to appear in the neighborhood. Parcels of land were being built up by contractors who had obtained building permits. Ted realized that the tide of official-dom was coming his way. Before long he'd be

Foam caulking fills 1″ gaps

greeted by a building inspector asking, "Do you have a permit for this structure?"

Ted wanted to improve his buildings, but he worried about investing more time and money without some official sanction. His land—zoned for both agricultural and residential use—is secluded, even though the parcel lies only twenty minutes from a metropolitan area. Since it was unlikely that inspectors had already seen his creations, he submitted to the County a rough plot plan noting pre-existing agricultural structures: a "goatshed" and "storage building." The County wrote thanking Ted for the plot plan—and informed him of the increased valuation and taxes! But Ted was elated: "The acceptance of the plot plan in effect legalized the buildings' existence. Any future problems with the County would focus on the *use* of the structures, not their right to exist."

Now Ted began a few modifications: "I thought it would be a real challenge getting a septic permit for a goatshed, but the Building Department and the Health Department didn't talk to each other. Nobody asked me what I was going to attach the septic system *to*, so I got a septic permit with no house. I built a detached bathroom midway between the Palace and the goatshed and ran the outflow into the septic. I installed sinks and showers in the goatshed and Palace, but ran their outflow into a gray-water system I designed."

Ted got back to finishing the Palace. He sheathed the inside with 1″ x 8″ redwood acquired from a house marked for demolition. He noticed that his inattention to detail resulted in some one-inch gaps in the walls. Insects and wind entered the house freely. But Ted found just the product: "Polycel 100 is great! It's a polyurethane sealant that comes in a can like shaving cream. You spray it into the cracks, and it expands to fill them."

With the elements held at bay, Ted moved from the "goatshed" to the Palace. A museum of sorts, the Palace reflects Ted's eclectic tastes. Every photograph, stuffed animal, and piece of pottery has a story. From the ceiling hangs a 45-star United States flag from the old Hearst mansion in Los Angeles. (Ted installed a toilet for an old-timer in exchange for it.) Sitting on a vintage organ is a three-hundred-year-old Indian skull that Ted found by a creek running near his property. By the front door hangs a beautiful brass chime made by Paolo Soleri, whom Ted got to know in Arizona.

Outside, by the front door, stands a homemade cannon with a four-inch barrel. In an annual Fourth of July ritual Ted began in honor of the Bicentennial, "British" Ted shoots grapefruit 400 yards across the valley to a neighbor's house. Of course, the "colonist" neighbor shoots back, but Ted thinks he cheats. "After we score a few hits, he uses baseballs because he doesn't have the finesse to shoot a grapefruit 400 yards. The trick is to wrap just enough tape around the grapefruit so it won't come apart—

but will splatter when it hits."

On the Palace walls are a number of interesting photographs. In one, Ted and his neighbors are enjoying an Easter costume party in skimpy bathing suits, Santa Claus suits, gorilla outfits, suits of armor, gunny sacks—anything goes. Even the chickens are in "costume," dyed with blue, green, yellow, and pink food coloring. A banner above the photo proclaims, "Dye the Chicken, Not the Egg."

"If there were more than one Ted," says a neighbor, "we'd have to do something. But crazy ideas and events flow from him and keep the community alive. He's so likeable, it would take a cold-hearted person to turn him in. The County knows he's up here, but they won't respond unless they get a complaint."

When pressed about the possible legal hassles he might face from the County, Ted replied, "I've given them an out. The structures were reported as existing. I've been paying taxes on them, the sewage is handled correctly, and I'm remote enough not to bother other owners or affect their property values. Say the worst happens: someone dies in a fire in one of my places. The County will say they didn't know I'd changed the use and leave *me* holding the bag—which is the way it should be."

Before I left, Ted insisted I visit his strawberry patch irrigated by the gray-water system. Ted's invention is a very inexpensive filter system that uses a ten-gallon drum filled with straw and some clever plumbing. "I ought to apply for a patent on this system. It's a winner. Aren't the strawberries great?"

Ted was right on both counts.

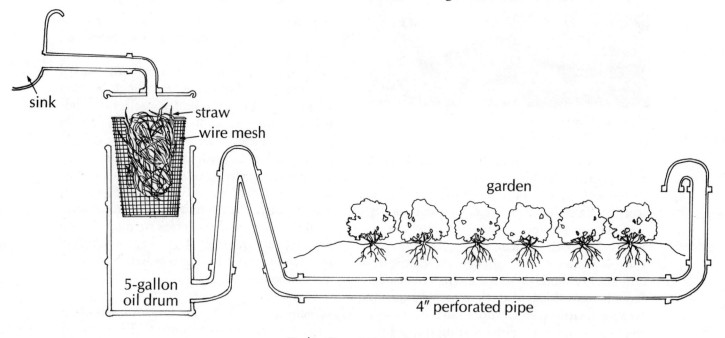

Ted's Gray-Water System

13. The Rise and Triumph of a Quasi-Legal Landlord

DURING MY travels, I met several owner-builders who had developed rental units for extra income. Most were operating in areas zoned R-1 (for single-family residential) or RA-1 (single-family residential with agricultural use). To me, it appeared that they ran the risk of losing their investment if officials discovered these zoning violations. But when Ed, a high school teacher, told me his story, I learned how a well-thought-out building and rental plan can generate good income and pay for itself in as little as six months, with little risk of financial failure.

"I was sick of giving a fourth of my income to a landlord. A friend had built his house with no previous experience, so I thought that was my best bet, especially since teachers don't make enough to buy a decent house." Ed bought

A house of brick and railroad ties

Placement of the posts depended on dimensions of the salvaged windows

seven acres of dense brush and woods zoned for single-family residential and agricultural use. "My father took one look at the property and said I'd made a big mistake." But spurred on by the thrill of land ownership, Ed was undeterred. "For the first few months, I spent my time clearing brush. Once the well was drilled, I started plans to build. I had two excellent building sites; one already had a small cement slab. I didn't have much money, so first I decided to build an inexpensive 700-square-foot house. Later I could get a loan to build a bigger structure using my first house as collateral. I designed a house, but applied for a permit for a residential workshop. A number of different structures—guest houses, art studios, barns, and workshops—can have all the conveniences of home and not be considered a primary structure as long as they have no kitchen. Some of the

code requirements are often relaxed if a structure is designated as something other than a primary dwelling."

He expanded the existing 18' x 20' concrete slab with a conventional T foundation, creating a split-level floor plan with 700 square feet of floor space. "I worked large pieces of granite collected from the property into the concrete as I poured. Probably half the foundation is rock, though none of it shows. Two strands of rebar run throughout, which keeps things pretty safe. The building inspector okayed it."

The framing is a combination of post-and-beam and conventional stud walls. For the post-and-beam framing, Ed used carefully selected 8″ x 8″ railroad ties, some new 8″ x 8″s, and several 8″ x 8″s scavenged from a beach after a severe winter storm. "Beaches and river mouths are great places to find good wood. Actually the

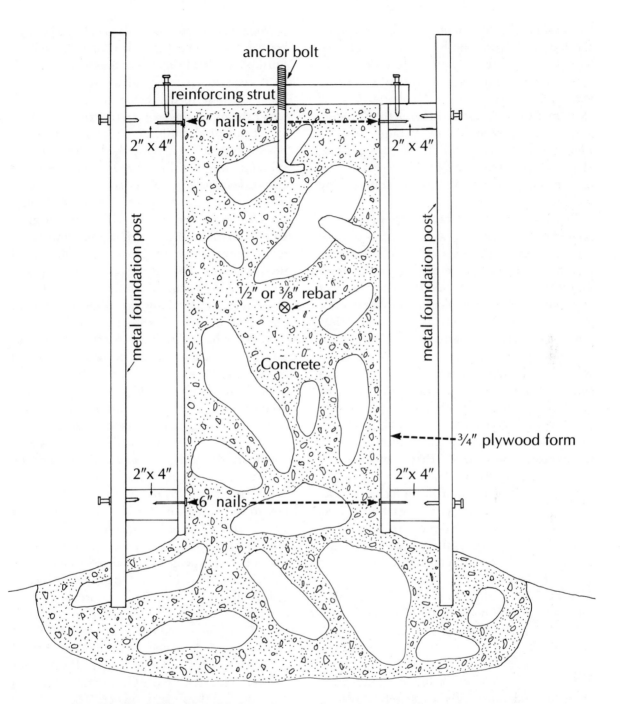

anchor bolt

reinforcing strut

6" nails

2" x 4"

2" x 4"

metal foundation post

metal foundation post

½" or ⅜" rebar

Concrete

¾" plywood form

2" x 4"

2" x 4"

6" nails

Conventional concrete T-foundation filled with stone saves up to 60% of the cost for concrete. To ensure good bonding, no rock should be wider than six inches (long, narrow ones are best) and should not touch each other or the side of the form. Place stone in layers as concrete is being poured. If some rocks are exposed after forms are pulled, use mortar to cover exposed surfaces.

twenty-foot 8″ x 8″s I found on a beach got me committed to post-and-beam, a type of construction I've always liked. The railroad ties cost about half as much as new wood. Make sure you don't get ties treated with creosote. Get the vacuum-treated or untreated kind; they're odorless." This post-and-beam framing method proved to decrease finish work considerably: "For example, two posts could serve as structural members as well as finished jambs for my windows."

Ed fastened the posts to the previously-poured concrete slabs in two different ways. For extra strength, the posts at the corners were placed in metal saddles secured to the concrete. Ed took great care in the location and plumb of each post because of its double duty as door or window jamb and structural component. The posts' tops were connected to the horizontal perimeter beam (or plate) by ¼″ thick metal straps lag, bolted in place. Once framing was complete, with doors and windows in place, he filled in the remaining spaces with mortared 8″ x 4″ x 16″ solid concrete/adobe brick. The concrete gave the bricks enough integrity to withstand heavy rains, while they remained soft enough to cut with a hand saw. Each brick takes up quite a bit of space, making for fast work. "Once the brick is laid, you have a finished exterior and the interior wall. You're done in one easy step." The brick, post, and beam wall costs less than $1.90 a square foot, about the cheapest code wall there is.

In the area built on a foundation, Ed used conventional methods: framed 2″ x 4″ stud walls, 2″ x 6″ subfloor, and an exterior sheathed in plywood.

For the beam ceiling, he used #3 grade subfloor material instead of the higher-grade, kiln-dried firs and pines. "By selectively choosing the wood, I came up with decent-looking stuff at less than half the cost of regular decking." The ceiling is covered with an inch of rigid foam insulation and a painted metal roof that should last longer and be more fire-resistant than other types of roofs.

Where he wanted interior paneling, Ed used bender board. Lumberyards classify his rough-finished redwood product as a garden material, so it costs much less than redwood paneling. Ed: "Allow it to dry before application so it won't shrink on the wall. In case there's shrinkage, I put black building paper over the studs before applying. This way the bright silver insulation won't shine through between the cracks, and the studs won't show."

The plumbing is copper pipe and ABS plastic. Plumbing had not existed in the original slab foundation, so Ed set all the necessary bathroom plumbing before pouring a small slab off the main slab. The kitchen is on the opposite side of the large pad. Ed ran the kitchen plumbing underground along the edge of the pad, joining it with the bathroom plumbing. The septic tank had to be in an inaccessible area, so Ed chose a fiberglass model that could be carried by two men.

Ed decided to build all his buildings to code, even if he had to interpret some of it rather creatively. At first, the workshop was the only structure on his property, and because it wasn't a primary dwelling, the local power company wasn't obligated to bring in electricity. It would be costly to lay heavy wire from the meter box to the other building site, over 100 yards away. But if Ed could somehow finagle *two* meters on the property, the power company would take up the expense and the total amount of electricity on the property would be doubled.

Initially, County officials assured him that two meters could not be permitted on one RA-1 lot. "They weren't convincing," says Ed, "so I asked around. Sure enough, a rancher showed me a way. I applied for an electrical permit for my well—something the County must grant to an agricultural property owner. The well was very near the workshop, so once the power was hooked up and inspected at the well head, I got another electrical permit for bringing power into the workshop. Because the original meter was for the well, I could legally install a second meter at the end of my property where I wanted

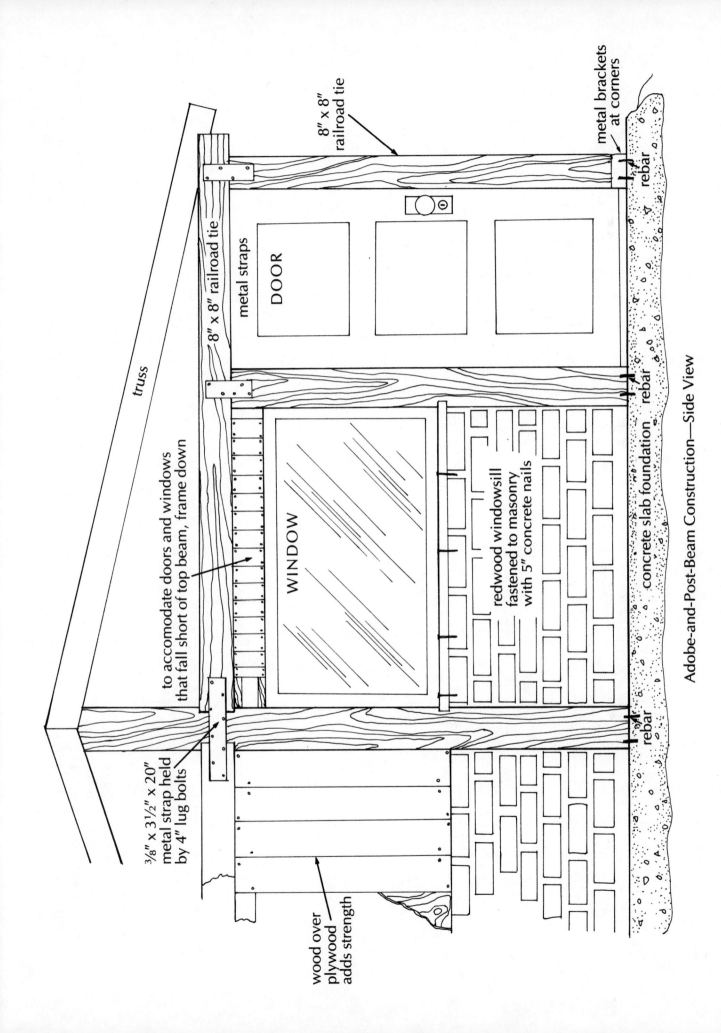

metal brackets at corners

rebar

8" x 8" railroad tie

metal straps

8" x 8" railroad tie

truss

DOOR

to accomodate doors and windows that fall short of top beam, frame down

WINDOW

redwood windowsill fastened to masonry with 5" concrete nails

rebar

concrete slab foundation

rebar

³⁄₈" x 3¹⁄₂" x 20" metal strap held by 4" lug bolts

wood over plywood adds strength

Adobe-and-Post-Beam Construction—Side View

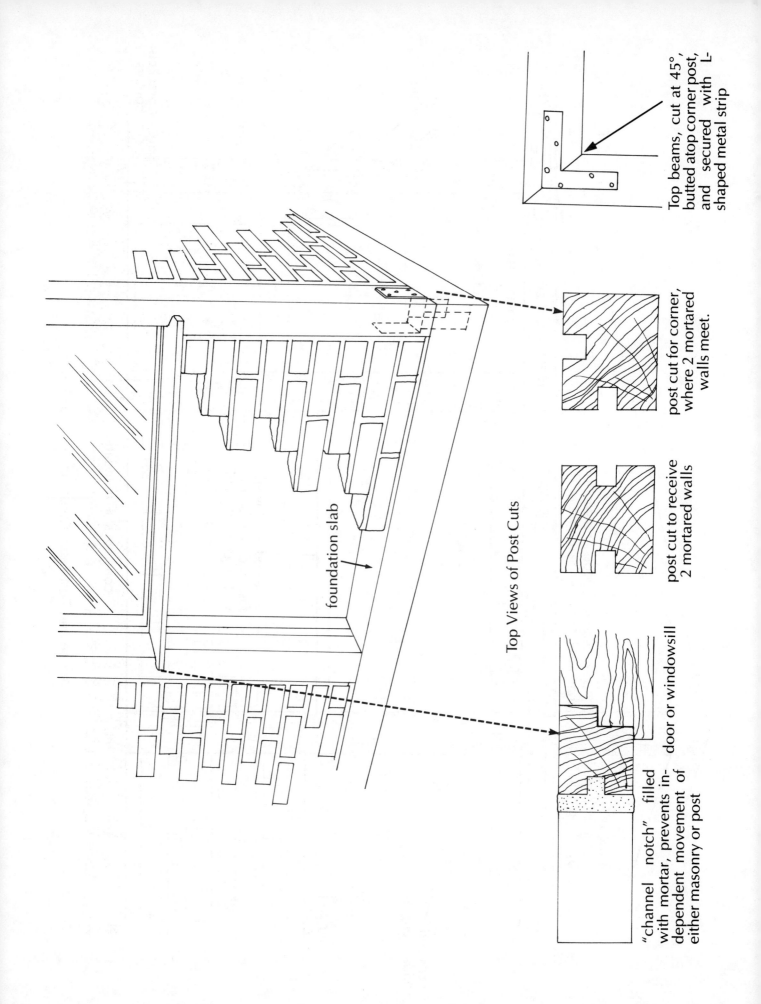

Top beams, cut at 45°, butted atop corner post, and secured with L-shaped metal strip

post cut for corner, where 2 mortared walls meet.

post cut to receive 2 mortared walls

Top Views of Post Cuts

foundation slab

"channel notch" filled with mortar, prevents in-dependent movement of either masonry or post

door or windowsill

to build my primary dwelling. If I hadn't used the trick with the well, I would never have gotten the power for the workshop."

Ed lived in the workshop for a year. It increased the value of the property and was set up to be changed into a house with the addition of a permanent stove. Using the workshop (which cost him $6,500) as collateral, he was able to procure a $20,000 loan.

Ed's next home incorporated many of the money-saving methods he used to build the workshop. The house did have a few embellishments: sunken bathtubs, a full kitchen—plus a lean-to greenhouse that served as a passive solar collector, a winter garden, and source of cleverly-funneled warm air. With tile floors, beam ceilings, three bedrooms, two bathrooms, and rich wood paneling, the house cost $25,000 to complete. Ed then took out a $30,000 mortgage, paid off the workshop loan, and had money left over for his next project.

The next summer found Ed building a 250-square-foot art studio with all the necessary permits. His main house had an extra-large septic tank, so the disposal from the studio ran into the existing system. The studio is nicely built, but doesn't have the handcrafted beauty of his other creations. "When I started the studio, I designed for rapid construction. I poured a slab foundation, with plumbing, framed with 2″ x 4″ studs, brought in a few circuits from the primary structure, sheathed the exterior with plywood siding, and covered the interior with sheetrock. Exposed ceiling beams supported # 3-fir decking material, and a combination of tile and carpet finished the floor. By the time I finished the studio, I had three houses on property zoned for one, all of them to code. As long as there are no permanent kitchens in a secondary structure, you're safe. Of course the zoning forbids a lot split, so you can't sell the structures separately. But you can rent them."

Ed researched the County rental ordinances and found out some interesting points: "A

Platform bed with storage space underneath

single-family residence can have five unrelated adults living in it, or two unrelated adults living with a family. Secondary structures can be lived in permanently as long as the person isn't cooking or paying rent. We keep the rent low to make the renters sympathize with us instead of the County. The renters sign a contract that states they are renting in the primary house, which keeps us pretty legal on paper. The art studio rents for $180 and the workshop rents for $300; the total easily covers the monthly mortgage payment on the entire property."

Ed feels his solution to the problems of unauthorized rentals is nearly foolproof: "I got in a heated dispute with a neighbor whose dog had killed one of our sheep. Rather than pay for the dead sheep, he phoned the County and reported that I was illegally renting secondary structures. When the inspectors arrived, we showed them the rental contract and the buildings, none of which had been altered since their completion. The plug-in kitchen appliances had been hidden in the cupboards, and we claimed everyone used the out-buildings at various times. The County would have had to do bed checks to disprove us, and they simply weren't interested. Basically, if you can show the officials something plausible, they'll leave you alone."

14. A Nomad Comes to Rest

WHAT IS a yurt doing in 20th century America? What *is* a yurt, anyway, and how did Stan Giovanni from Brooklyn end up with a round house that resembles a green-mantled mushroom?

The first yurts were designed for a nomadic lifestyle—lightweight, easy to assemble and dismantle. Round houses of skin and poles bound together by woven rope, they sheltered such Asian notables as Attila the Hun and Genghis Khan. Yurts are still widely used by today's Mongolian herdsmen. Now a handful of American builders have adapted the basic principles of this centuries-old dwelling to wood. Stan: "I didn't know what a yurt was until I happened by an alternative structures exhibition. A friend

The wooden yurt

94

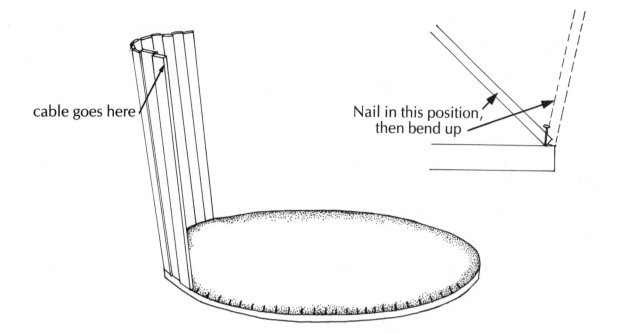

cable goes here

Nail in this position, then bend up

who was with me kept saying, 'It looks like a Martian spaceship.' But for me, it was love at first sight."

Stan owned a couple of acres in the country and thought a yurt would be just the thing for country living. "So I bought two: one eighteen feet in diameter with 250 square feet, and one thirty feet across with 700 square feet. The land cost $30,000 and the yurts $18,000. For 1980, that was a good deal."

A truck delivered Stan's yurts in modular sections, lightweight enough for two men to carry. A yurt can sit on any foundation. Stan chose pier-and-post because it was the least expensive and the easiest to build. He installed a conventional floor joist system and a ⅝″ plywood subfloor. It took three men three days to assemble both yurts into waterproof shells. Stan completed the plumbing, electrical wiring, and finish work on weekends.

If a yurt's basic lines do for you what they did for Stan, you can buy a modular kit from one of a small number of manufacturers. Or you can build one in your garage and assemble it on a weekend with the help of a few friends. A yurt's

Vertical door in outward-leaning wall

95

Top View of Wall

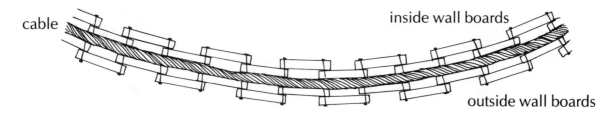

cable   inside wall boards

outside wall boards

shape makes it ideal for simple modular construction. Today's wooden yurts, like their skin and pole predecessors, are lightweight. The modular wall sections are framed with 2″ x 4″s. And even with a thirty-foot ceiling span, the largest rafter is a 2″ x 6″.

Only two shapes are needed to construct the walls and roof. First, a series of identical, wedge-shaped trapezoids are raised along the circumference of the subfloor. Once the "wall" circle is complete, a cable strung around the top of the wall is tightened at a turnbuckle. For the roof, a series of triangles are squeezed into an O-shaped compression ring, which does double duty as the frame for a center skylight. Nails attach the circular wall to the subfloor and the

The foundation: cross-braced beams set in concrete

roof, but essentially, the structure is held together by the two cables that draw the building in against itself.

When you want to install cabinets, closets, and countertops inside, the yurt's circular outward-leaning walls present some interesting challenges. Stan: "Most kitchen and bathroom cabinets are built to fit straight lines and square corners. I didn't have the skill to modify regular cabinets or build special cabinets, so I built some conventional interior walls and hung standard cabinets on them."

Stan claims the outward-lean of the walls has advantages: "A curved bench along the wall has a naturally comfortable backrest. Bookshelves attached to the upper half of the wall don't take away valuable floor space. The outward-leaning

windows collect less outside dirt and shed water more readily than vertical windows.

"Some guests say the outward-leaning walls disorient them a little. I felt that way too in the beginning, especially after a couple of beers. But now that I've gotten used to it, a traditional, box-like room looks too stilted to me."

To make sure his yurt was energy-efficient, Stan layered the roof from bottom to top with one-inch pine, one-inch rigid insulation, ⅜" plywood, hot mopped tar, and four inches of sod. He fully insulated the walls, double-paned the windows, and stapled six inches of fiberglass insulation underneath the subfloor.

In a region that commonly sees 20° F. nights, Stan's large yurt is easily heated by a small wood-burning stove. "But if I were to do

Turnbuckle on the rafter cable that holds the wall together

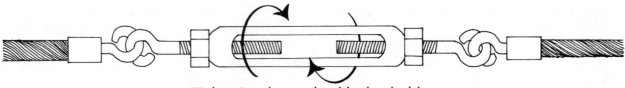

Tightening the turnbuckle that holds
the outer walls together

it over again," he confides, "I wouldn't buy a kit. I'd make my own patterns, one for the walls and one for the roof. Instead of watching television a couple of hours a night, I'd build the sections until the parts of the pie were ready to put together. Anyone can build a yurt in their garage and take it to a building site and assemble it in a couple of days. I don't know why more people aren't excited about yurts. They're inexpensive, fun, and practical."

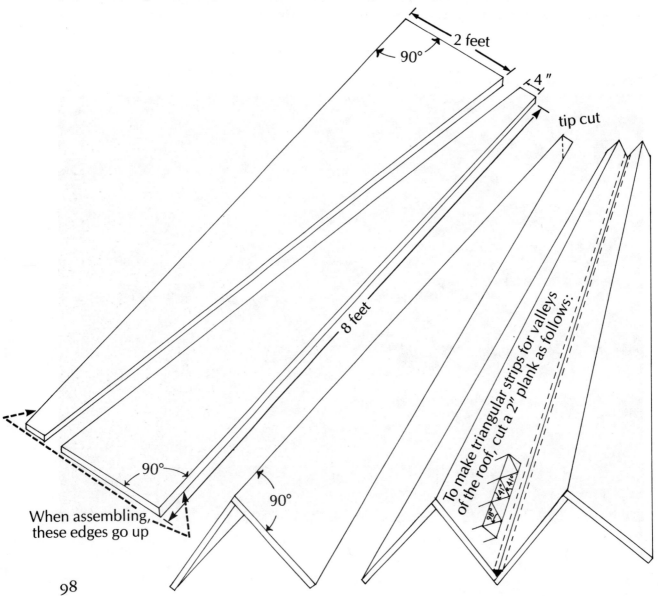

2 feet

90°

4 "

tip cut

8 feet

90°

90°

When assembling,
these edges go up

To make triangular strips for valleys
of the roof, cut a 2" plank as follows:

15. The Treehouse
in the Cathedral

TO SAY merely that Alfred is an eccentric who lives in a treehouse doesn't begin to explain a lifestyle that offers a refreshing perspective on what is an adequate home.

A big, powerful man with a cascading golden beard, Alfred—on first glance—might be seen as some sort of counter-culture holdover from the sixties who is stationed in poverty because of lack of skill and/or employment. Not so: Alfred could've taken the conventional route. As a former Pac/Ten football star and highly skilled craftsman whose remodeled houses have been featured in *Sunset* Magazine, Alfred has proven he has the education and talent to compete with the Joneses. His lifestyle is a philosophical commitment: "We Americans've grown into taking

"Slip" foundation, seen from below

The treehouse

Cruz Mountains. He sleeps and cooks (that is, if he isn't entertaining guests around the campfire circle) in an expertly-made treehouse. To Alfred, the treehouse is incidental; he calls it his "dry space."

Alfred would not accept the idea of a lifetime mortgage: "In 1976, this undeveloped parcel cost $25,000. If I hadn't touched it, it would be worth about $40,000 today. When I bought it, I made a commitment that I'd never incur another debt. Now I've paid it off and have no mortgage hanging over my head."

To many prospective buyers, Alfred's heavily-wooded parcel would have appeared depressing, with too little sunlight. Alfred saw it differently: "I knew I could change the picture with a chainsaw and a lot of work. It took me four years of weekends and days-off to get the open feeling I wanted." After he selectively removed some trees and limbed others, the inhospitable forest began to resemble a park. A huge original-growth redwood, ten feet in diameter, was isolated so its massive and towering beauty could be seen.

Alfred's meticulous beautification program wasn't for the faint-hearted. Some of the limbing required climbing nearly 200 feet into trees. When removing a tree he burned out the stump and covered it with dirt so no trace was left. He brought in new soil and organic fertilizers, planted a citrus orchard with sixteen varieties represented, a small deciduous orchard, and colorful plants like rhododendrons, azaleas, camellias, and daffodils. Meadows were sowed, and wildflowers sprang up. What had been a tangled forest became a dazzling array of isolated stands, specimen trees and colorful exotics.

"The main thing," Alfred insists, "is getting a piece of land you can fall in love with. It should have natural beauty, privacy, and if possible, naturally produced water. Develop the piece aesthetically first, then think of ways to get a place to stay dry."

During the time he prepared his natural cathedral, Alfred lived in two six-man tents:

too much and thinking we need it. I like to think I haven't disfigured the landscape or created an energy-guzzling dinosaur [*modern multi-appliance house*]. I've forsaken some of the so-called conveniences, but I've also forsaken a lifetime of debt. I have control over my life. I don't have to support an army of gadgets like dishwashers and the rest of it."

Alfred's approach to making a "nest" is to celebrate the natural beauty around him and to build a living space no bigger than necessary. Consequently, Alfred lives on a manicured five-acre parcel in a secluded hideaway in the Santa

one for him, the other for his valuables. Not until he had finished working his forest into shape did he decide to live in a tree. "I wanted to be in the middle of the beauty and have a view of the coast. To get the height I needed, I had to build a treehouse. I picked a large tan oak with three trunks because it was suited to support a large platform."

There was an engineering problem: in heavy winds, each trunk moved separately. If the treehouse was secured rigidly to all three trunks, continuous wind action would tear it apart. "First I trimmed the tree to lessen the wind drag. Then I decided on what I call a 'slip foundation.' I notched each trunk and placed a lintel in each notch. Each lintel was leveled to the others and secured to respective trunks with wooden pegs. I fashioned a huge triangle out of

The kitchen "area"

heavy beams and placed it on the lintels. This way, the three trunks became united to support the treehouse. And because the triangular frame isn't attached to the lintels, each trunk can move independently under the frame."

Treehouse skylights, as seen from above

Building a Slip Foundation for a Treehouse

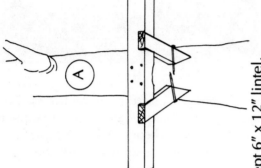

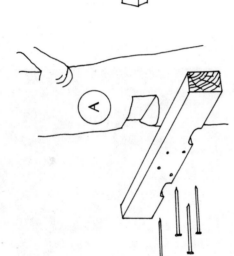

Step 1: Use taut string and a level to locate placement points for each lintel.

Step 2: Notch trunk A to accept 6" x 12" lintel. Before affixing with 14" spikes, tar notch to prevent decay to tree. Brace lintel with diagonal braces, as shown.

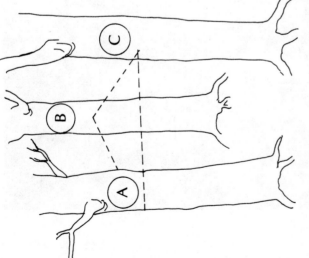

Step 3: Notch trunk B to accept longer lintel. Insert brace in notch in trunk C to hold other end of lintel.

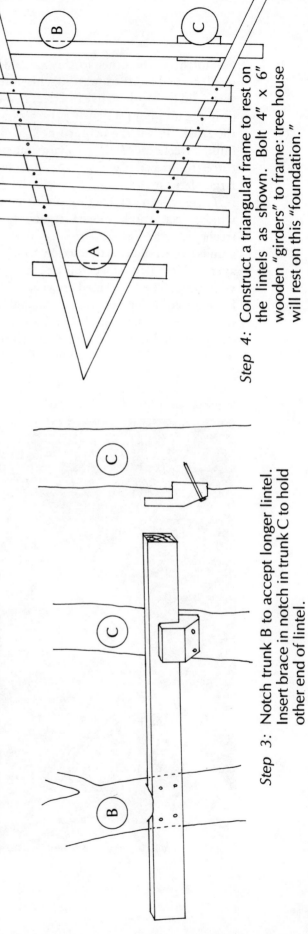

Step 4: Construct a triangular frame to rest on the lintels as shown. Bolt 4" x 6" wooden "girders" to frame: tree house will rest on this "foundation."

Sleeping area, with kerosene lamp

Alfred's slip foundation has survived sixty-mile-per-hour winds. On it sits a cozy little cabin made entirely of salvaged lumber. The structure is square, true, and absolutely waterproof. It even has a shingle roof, skylight, and small porch.

Alfred has no plans to bring in electric power. At the high end of the property, a spring that produces five gallons a minute, twenty-four hours a day, is guided into a gravity-powered water system that flows downhill, requiring no pump to operate.

The inspiration for Alfred's gravity water system came from the ancient Romans and what Alfred calls his "aquarium habit." "As a teenager, I read a book about Pompei, where most houses and fountains had gravity-fed water systems. Also, I used to be an aquarium freak. I

had five one-hundred gallon tanks and was always draining one into an outside garden with a hose or plastic tubing."

Alfred's water system, over fifteen hundred feet in all, wanders into ravines, through dense brush, around rock outcroppings, and under giant redwood logs. Every 100 feet or so, a smaller pipe branches off the main. To figure out where to lay the pipe, Alfred first drew a rough topographical map of the property. "The key is to follow the contours, rather than to work in straight lines. Also, it's important to keep your main line as high as possible so you'll have greater flexibility in adding secondary lines."

Before laying the actual pipe, Alfred laid out the whole system in garden hoses he'd bought and borrowed. "Starting at the spring, I connected the hoses and laid them out following the topo map. Because of the vegetation and constant ups and downs it was often difficult to judge where to lay the hose. To help locate downhill, I used a sight level [*a device resembling a spyglass, with a bubble level on top. A string level, builder's level, or even a low-power telescope with a level taped on top could be used*]. The sight level allowed me to look along a rise 200 yards away until I found the lowest point. I had to carve down some high areas and reroute lines to establish the flow I wanted. Once the hose system worked, I replaced it with buried plastic pipe. In parts of the system, I have up to forty pounds pressure."

Even though the treehouse is thirty feet in the air, it is still below the spring which lies 150 yards away on a ridge. In fact, Alfred cleverly calculated the treehouse's elevation to make sure it would not exceed the source of water.

For lighting, Alfred uses flashlights and kerosene lamps. "What was good enough for Abraham Lincoln," he grins, "is good enough for me." With a kitchen resembling a small yacht's galley, hot water is produced by wood heat, a copper coil, and a small storage tank. At ground level Alfred has an outhouse that he

Another view of the kitchen and overhead skylights

moves every month or so, "to spread the richness throughout the land." For bathing and soaking, there are a solar-heated shower and hot tub.

Surely, Alfred must tire from his back-to-the-basics lifestyle? "Nope. I've been here six years, and I'll be here the rest of my life. Perhaps some day I'll move back to the ground and build a little stone house. But to do that, I'd have to be committed to family living."

What about the structural soundness of the tan oak's foundation which, after all, would never be approved by a building department? "This tree has been here over one hundred years. That's longer than most manmade foundations last." As for the building inspector, Alfred adopts a devilish grin and an alibi: "If an inspector happens by, I'd explain I built this treehouse for my nephew who visits every summer, and the property is for recreational use. My biggest worry is having some clumsy friend fall off the ladder. In fact, the ladder has turned out to be a good screening device. I have decided women who won't climb the ladder aren't my type."

16.   Willpower

WITH NO building experience, Paul Schurch had chosen a building site developers had passed over. Because of the immense granite outcropping that dominated the steep landscape, neighbors dubbed Paul's lot "Boulder Gulch."

Clearly the odds were not on his side, but desperation had pushed Paul into this gamble. "By 1980, most of the good building sites in the area were taken. I couldn't afford to buy the type of house I wanted. I knew I was taking a chance, but I wasn't getting any younger, and I really didn't see any alternatives." The parcel cost him $50,000—about $20,000 less than the surrounding lots.

Paul credits his father's approach towards life with giving him the determination to build his own home. "My father flew experimental air-

Nestling among boulders

craft. I learned that it's possible to do things other people won't even attempt. In my family, we'd take on difficult tasks just to see if we could succeed. Also, as a teenager, I worked for an organ maker in Switzerland. From him I learned that to make something truly beautiful, you need patience and attention to detail.

When Paul submitted plans to the County building department, "they were rejected. I asked to talk to the engineer who found them unacceptable, and he showed me what my mistakes were. My corrected plans were okayed."

Because of the boulders in Paul's plot, "building a conventional foundation was out of the question. A conventional floor plan, with interlocking rectangles, was also unworkable. Instead, I let the placement of the foundation footings influence the shape of the house. That's why the rooms are trapezoids—and why there's a boulder in the bathroom."

Paul was able to establish some footings in the dirt; other footings had to rest on boulders. "Before I tied into a boulder, I checked it for absence of large cracks and suitable size and stability. Then I drilled into the rock and secured expansion bolts so the footing and rock would be united with steel."

Attaching beams cut from old bridge timbers to the footings, Paul built a conventional 2″ x 10″ joist system, 2″ x 4″ stud walls, and a beam ceiling. "Framing is easy. From the County I got a span sheet that tells you exactly how big and how far apart beams must be for different spans."

Paul tried to cut costs by using recycled lumber, but wasn't always successful: "The re-

Kitchen area, directly off the bedroom

The sunken bathtub

cycled big beams were fine, but the 2″ x 4″s were usually more trouble than they were worth. They were case-hardened, which made the warped ones impossible to straighten and almost impossible to nail into them. I ended up using a lot of them for firewood. I would have been better off with new pliable wood."

Paul had no major problems wiring his house. "I read *The Electrician's Bible* in one evening and finished the rough wiring the next day. Except for a couple of three-way switches that a friend helped me with, the finished wir-

ing was easy too." But plumbing had its pitfalls: "Even thinking about plumbing makes me grind my teeth. First, I tried to use second-hand fixtures. *Never* buy an old toilet with parts missing. Don't even accept one as a gift! By the time you've run around to fifty plumbing departments, always getting a piece that'll probably do the job, you could have paid for a new toilet. Then, when I finished the rough plumbing, I discovered I'd seriously miscalculated: the septic tank and leach field ended up ten feet higher than my lowest drain. To push the toilet

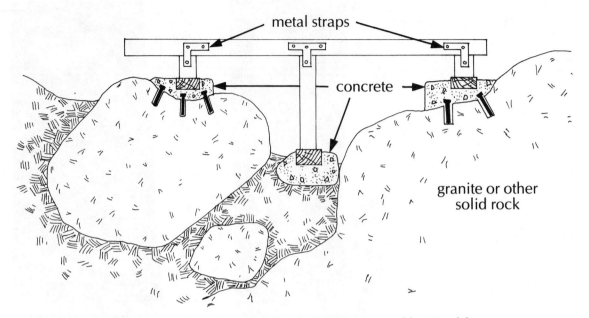

metal straps

concrete

granite or other
solid rock

Schematic Drawing of Foundation Supported by Boulders

How the boulder enters the bathroom: exterior window
is footed in a crevice carved in the rock

discharge uphill, I had to install a thousand-dollar sewage pump. The rest of the drains lead to a gray-water system that flows into the grass below the house. My hope is to keep the grass green year 'round so it'll act as a buffer against brush fires."

Paul's attention to detail and good taste has created a striking home. The living area is finished in perfectly taped, untextured sheetrock. The picture window is flanked by two small leaded and beveled glass windows. For heating, he installed an airtight Fisher woodstove. For hot water he uses a solar "breadbox" system in the summer and a wood-burning water heater in the winter. The flue pipes from the water heater and wood stove merge artistically before disappearing through the ceiling.

Two steps up from the living area is the kitchen. Paul bought a reconditioned 1930s gas range and had custom cabinets made. He added a small stained glass window above the kitchen sink and a Dutch door to the outside. But his most innovative work is in the bathroom, where he suspended a standard free-standing cast iron tub through a hole cut in the floor and sur-

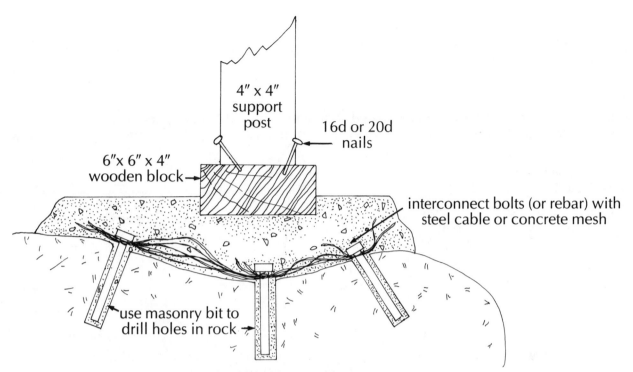

4" x 4" support post

16d or 20d nails

6" x 6" x 4" wooden block

interconnect bolts (or rebar) with steel cable or concrete mesh

use masonry bit to drill holes in rock

Close-Up of Individual Support Post Footing

Solar "breadbox" water heater

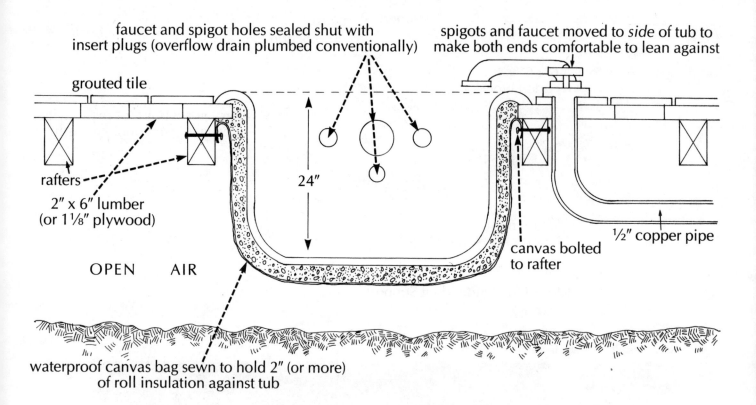

faucet and spigot holes sealed shut with
insert plugs (overflow drain plumbed conventionally)

spigots and faucet moved to *side* of tub to
make both ends comfortable to lean against

grouted tile

rafters

2″ x 6″ lumber
(or 1⅛″ plywood)

24″

OPEN   AIR

½″ copper pipe

canvas bolted
to rafter

waterproof canvas bag sewn to hold 2″ (or more)
of roll insulation against tub

Sinking a Cast-Iron Bathtub
(tub hangs from its lip, through hole cut in sub-floor)

rounded the sunken tub with floor-to-ceiling, wooden-framed windows. Paul bathes indoors while appreciating the outdoors.

The inclusion of a boulder into the room is interesting and attractive. "Instead of avoiding the rock, I decided to incorporate it. By involving the boulder, the house and the earth become united." Paul carefully cut each hardwood floorboard to fit tightly against the rock. With a masonry blade, he scarred a line in the boulder to make a seat for one of the windows.

Paul altered the tub's plumbing by removing the fixtures from the end of the tub and placing

them along the side. "With the fixtures out of the way, two people can lean against the ends without one of them getting skewered." Underneath the bathroom, where the tub would be exposed to the elements, Paul fashioned an insulated, fiberglass-lined canvas envelope for the tub and stapled it to the subfloor.

When Paul finished the bathroom, he was exhausted and out of money. "To end things on an economical note, I sheathed the house in ⅜″ plywood, stuccoed the exterior walls, and applied rolled roofing to the roof. All told, including the lot, I spent $60,000 for a 600-square-

Wood-burning water heater

Converging flues
from wood stove and water heater

foot house I'm really proud of."

Paul encourages others to build their homes: "Anyone can do it—if they're overconfident to begin with, and have the willpower to see the thing through once they realize they're in over their heads."

17. A Three-Story Tower Cottage

DICK SAYS that anyone can build a thirty-eight foot water-tower cottage: "It just takes careful planning and a lot of work." But Dick is very modest and doesn't like to mention all the ingenuity, creativity, and resourcefulness he put into his three-story, 700-square-foot dwelling with a 3,000-gallon water tank on top. The whole project cost less than $10,000.

As Dick remembers it, the idea sprang from sketches he made during a long, wet winter. "My wife and I had talked about making a large storage tank with a gravity-fed system for the garden and fire protection for our home. One night I drew a tank tower. I doodled a bit and drew in some windows. It looked like a large Dutch windmill without the blades."

A few weeks later, Dick traded his drafting skills for four 30-foot telephone poles. "I always thought that treated telephone poles would

The three-story cottage

Rebar railing and Irish lamppost

The central tower, topped with water tower

make an ideal structural material. The poles provided the catalyst. I envisioned them as the four corner posts for the tower, and after I got them home, there was no stopping me. I started looking around for more inexpensive structural material.

"Not too far from my home, a salvage company had removed most of an abandoned train trestle. They left behind the most inaccessible members. I climbed down into the canyon with some tools and dismantled what remained of the trestle. All of the wood had been treated to withstand termites and dry rot. The canyon

walls were too steep to carry out the timbers, so the next weekend a friend and I hoisted them 300 feet with a block and tackle. By the day's end, we had winched up fifteen 16-foot 6″ x 8″s. This was all the wood I needed to make the platform for the water tank and the joist systems for the second and third floors."

Once Dick had gathered the structural wood, he began work on the foundation. "This is earthquake country, and to support twelve tons of water thirty feet in the air, I made sure the foundation was overbuilt. I designed a square slab, fifteen feet on a side and 30 inches deep. At

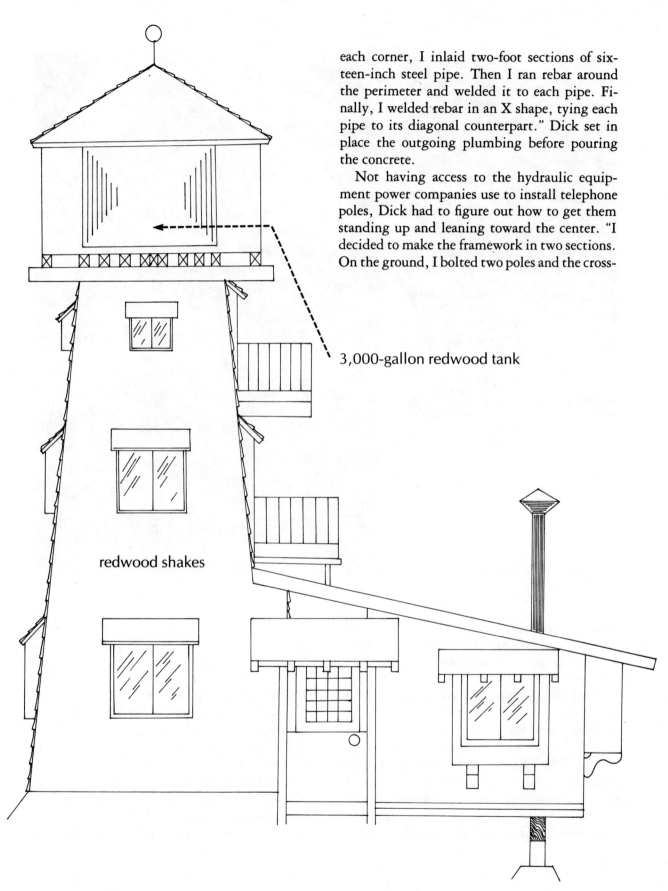

each corner, I inlaid two-foot sections of sixteen-inch steel pipe. Then I ran rebar around the perimeter and welded it to each pipe. Finally, I welded rebar in an X shape, tying each pipe to its diagonal counterpart." Dick set in place the outgoing plumbing before pouring the concrete.

Not having access to the hydraulic equipment power companies use to install telephone poles, Dick had to figure out how to get them standing up and leaning toward the center. "I decided to make the framework in two sections. On the ground, I bolted two poles and the cross-

3,000-gallon redwood tank

redwood shakes

pieces into a flat-top A shape. Using a block and tackle high in a nearby tree, I was able to rotate the section into two pipes at about an 80° angle. The second section presented a bigger problem. Since it had to lean toward the first section, I had to rotate it to 90°, and then drop it over another 10°." Once the two sections were in place, crosspieces were bolted in—at which point, the structure looked somewhat like an oil derrick. With the super-structure set solidly in place, Dick poured hot tar into the pipes around the base of each pole to protect against moisture and insects.

Using the salvaged trestle material, Dick framed in the second- and third-floor joist systems. Later he covered the joist systems with 2″ x 8″ tongue and groove, kiln-dried pine—

which served as flooring above and ceiling below.

To finish the wall framing, Dick used conventional 2″ x 4″ framing and tacked on ½″ plywood for sheer strength. On top of the structure, he built a heavy-duty girder system to hold up the 25,000 pounds of water, covering the girders with 2″ x 6″ redwood and tar.

Dick covered the exterior of the structure with taper-sawn redwood shakes (the least expensive redwood shake), using aluminum nails that won't bleed and stain the wood. "I rigged up a scaffolding, with the ropes secured to the top of the tower, and pulled the platform up and down, like a window washer does on a skyscraper."

Now Dick was looking for ways to get from

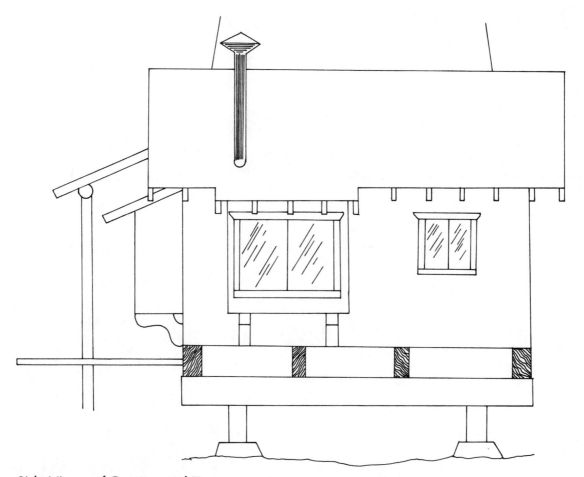

Side Views of Cottage and Tower

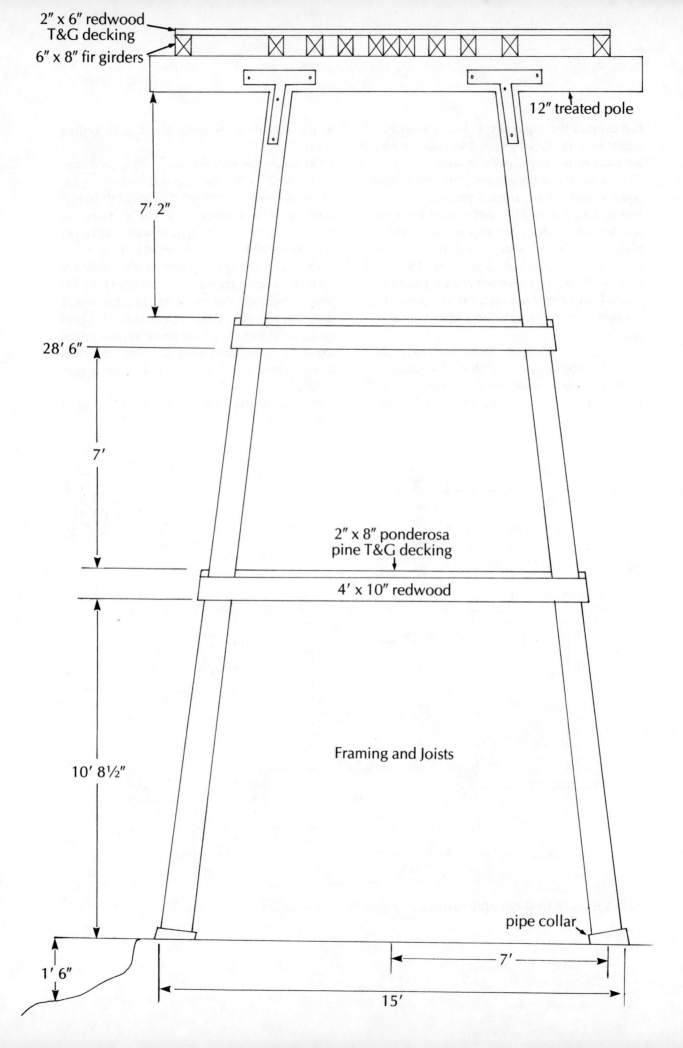

2″ x 6″ redwood
T&G decking

6″ x 8″ fir girders

12″ treated pole

7′ 2″

28′ 6″

7′

2″ x 8″ ponderosa
pine T&G decking

4′ x 10″ redwood

Framing and Joists

10′ 8½″

pipe collar

1′ 6″

7′

15′

floor to floor: "I was faced with the problem of building an inexpensive staircase that wouldn't take up valuable floor space. Years ago I took a welding class, so I decided to put the knowledge to work." Dick bought most of the material for the staircase at the local metal salvage yard. For a center post, he cemented a 2½" pipe into the ground 3½ feet from the tower. He cut a slightly larger pipe into seven-inch sections (seven inches is the standard rise per step). To each section he welded a 32" long channel support for the 2" redwood step, and slid each section/step onto the center post. When the steps reached the level of the second floor, he fanned them out evenly and welded them into place.

For the railing posts, Dick used #8 and #10 rebar and 1" steel cable for the handrail. To reach the third floor, Dick welded up a rebar ladder and fastened it to the exterior wall.

Rebar ladder leading to the third floor

First-floor ceiling, showing beams and bolts

Tar-filled pipe collar protects telephone pole from moisture and insects

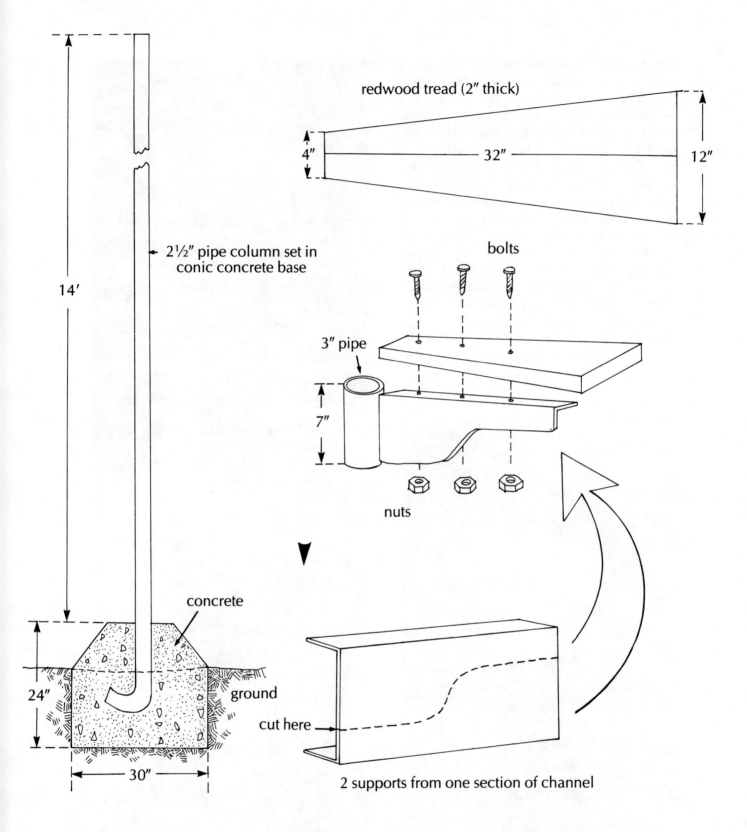

14'

2½" pipe column set in
conic concrete base

redwood tread (2" thick)

4"

32"

12"

bolts

3" pipe

7"

nuts

concrete

24"

ground

30"

cut here

2 supports from one section of channel

Building the Circular Staircase

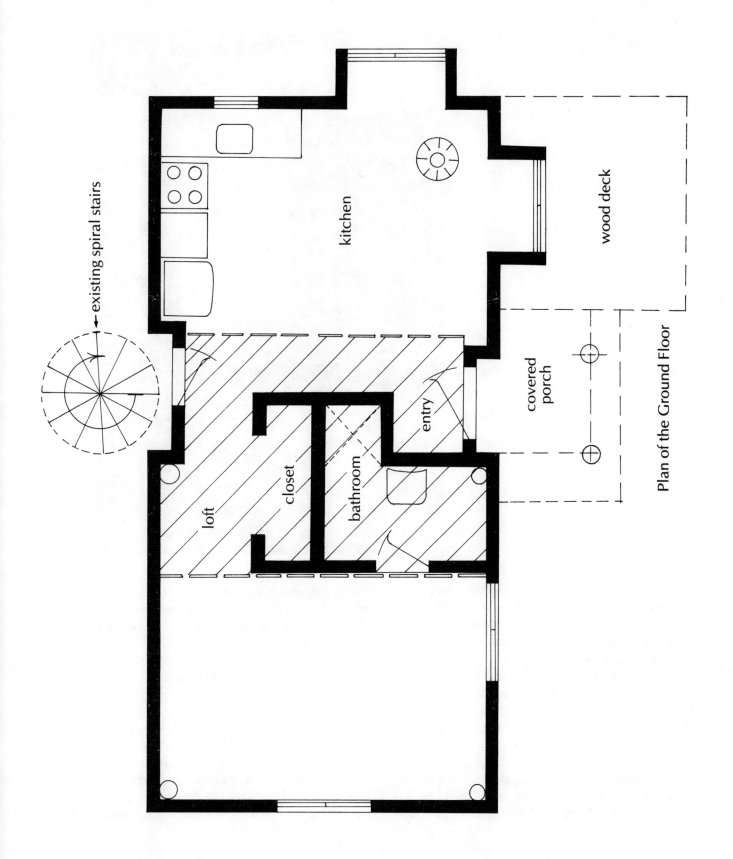

existing spiral stairs

kitchen

wood deck

covered porch

loft

closet

bathroom

entry

Plan of the Ground Floor

Third-floor bedroom (bed folds up against wall when not in use)

About the time Dick finished the stairs, he decided he needed more floor space. The conventionally-built square addition, 15′ x 15′, gives him 450 square feet on the ground floor—enough space for a living room, kitchen area, bedroom with loft, and bathroom. A skylight provides the addition with natural lighting.

In the "kitchen" corner of the living room, Dick fitted a small propane gas range, sink, and refrigerator. But while installing a fiberglass shower, sink, and toilet in the bathroom, he discovered he'd made a mistake: "I had put the toilet drain too close to the wall. The toilet fit OK, but not the water closet. Since all the out-going plumbing was imbedded in concrete, I had to improvise: I attached the water closet to the wall *above* the toilet and connected them with ABS pipe."

To provide lighting and circuitry on all three floors, Dick ran two 220-amp circuits into the tower from his main house. The entire tower is heated with an efficient free-standing wood burning stove that Dick designed and built himself. "I cut the middle section out of a discarded propane tank and welded the two ends together, cut out two doors, and welded on some hinges. I cut a hole in the top and welded up a flange to attach the stovepipe. On the bot-

120

Third-floor ceiling supports the water tower (note chains for bed)

tom, I welded the stove's stand—a pipe with a plow disc for the base." The fireplace works so well that Dick made a larger version for his 1,500-square-foot house. He's even built several for friends and neighbors.

Dick fixed up the 12′ x 12′ second story of the tower into a study with drafting table and desk. The third floor is outfitted with a desk and bed that can be folded up into the wall.

When he finally got around to placing the 3,000-gallon redwood tank on top of the tower, he turned back to his trusty block and tackle: "I rigged the block and tackle to a tree, hoisted up the parts of the tank, and assembled the whole thing on the top with my wife's help."

Dick has plenty of encouragement for others: "If you're willing to work hard, you can build very inexpensively. The key is to look around for discarded materials. It's very satisfying to make something beautiful and useful with materials that someone else considered useless."

Although his family could rent the water-tower cottage, they prefer to use it as a guest house. "And someday, when my dad is ready to move in with us, he'll live in the tower. He knows this is his home."

18. The "Periscope House"

CAPTAIN JOHN SMITH is better remembered for founding Jamestown and for his relationship with Pocahontas, than for fathering do-it-yourself housing in colonial America.

In the fall of 1607, Smith arrived from England with over 600 fortune-seekers, most of them second and third sons of wealthy families—men unaccustomed to physical labor. One of Captain Smith's priorities was building adequate shelter before the harsh winter set in. With a handful of crude tools, Smith and his legion of unskilled gentlemen built wattle-and-daub houses—a form of construction that required nothing more than earth and straight-trunked trees.

The Jamestown settlers felled trees, squared logs with axes or adzes, and dragged them to the building site. Using post-and-beam framing held together with interlocking joinery and wooden pegs, the settlers constructed rectangular houses. To guard against lateral movement, they placed at least one diagonal brace in each wall and placed a series of steep-pitched roof trusses atop the walls. Across the trusses, workers attached a framework of saplings that provided extra stability and a base to which to tie tightly-bound bundles of thatch.

After the frame and thatched roof were in place, the workers started the wattle—a latticework of thin interwoven sticks, constructed in two stages. First, workers hand-fashioned dowling about about one inch in diameter and lodged it vertically about five inches apart until a completed section looked like jail bars. Then, flexible willow shoots were woven horizontally through the dowling and fixed to the posts. The

The ferrocement roof

resulting weave was plastered on both sides with mud (or daub) to form a reinforced earthen panel.

Smith's houses had few embellishments. Floors were usually earth. Smoke from the fires passed out a hole in the thatched roof or up a stick-and-mud chimney. Because settlers had no glass, they built shuttered openings to admit ventilation and natural light.

Today, wattle-and-daub houses have problems meeting minimum building code requirements. But a handful of builders have stuck to the time-consuming process introduced by Smith and his men. In remote areas, a scattered few living on paupers' budgets have turned to this style of construction, which has also been employed to build secondary structures and historic reproductions. New Yorker Greg Moon, who built the pure form for summer use, sees modifications as sacrilege: "Modifiers ought to be honest about it and build a plywood box. After all, wattle and daub is an art form as well as a structure."

But some other "modifiers" have updated the wattle-and-daub process, using strictly modern technology. The most common modifications include replacing thatch with modern roofing materials, axes or adzes with a chainsaw, wooden pegs and joinery with metal spikes, and in place of interwoven sticks plastered with mud, metal lath or chicken wire plastered with concrete mortar.

Children think they've stumbled onto a *Star Wars* hideaway and want to explore it. Adults grope in their memories to establish a

The exterior staircase

One of the windows, with its ferrocement "bonnet"

frame of reference, but can't find one. The structure has no ties with classical, Victorian, colonial, ranch, modern, or any other common style. It's a free-form sculpture.

Dubbed the "Periscope House" by a teenage passerby, Mary Gordon's studio shows the unlimited possibilities of ferrocement, invented in 1943 by Luigi Nervi, an Italian engineer. He found that correctly mixed mortar will stop steel from oxidizing, and that mortar in combination with reinforced mesh becomes waterproof, flexible, and extremely strong. It's less rigid than concrete, but actually much stronger than separate but equal amounts of steel or mortar. Mary: "It's an exciting medium that invites creativity. Ferrocement lets you break away from the straight lines and square corners traditional building materials dictate. Lack of imagination is the only real limitation to design possibilities."

When you see the space-age studio and then meet its designer, there seems to be an incongruity. Mary Gordon, a successful landscape architect whose work has appeared in *Sunset* Magazine, lives in a conventional wood house on a conventional, typically well-tended Palo Alto street. She served on the city's Planning Commission for 15 years and received the 1980 Citizen of the Year Award. Mary explains how she became involved with a nonconforming building medium: "We owned a lot in a coastal community, and wanted to build a studio for weekend and summer use. I had salvaged some French doors and thought a small shingled cottage would be nice. To get some ideas, I visited an architect friend and found him all excited about a ferrocement structure he had just designed. I was intrigued by the design possibilities of the flowing curves and went home to toy with different ideas, but found I couldn't get a feel for the finished product in two dimensions. For example, it's nearly impossible to draw the undulations of a reverse-curve, asymmetrical roofline. So instead of *drawing* ideas, I started sculpting clay models until I found what I wanted. My final design was definitely free-

form but did have somewhat of a Moorish look. The 'periscopes' were incorporated to get ocean views over the rooftops of the neighboring houses."

To satisfy County Building Department requirements, Mary drew structural plans, foundation plans, and side views. "At first, the County wanted computations for the stresses throughout the building—which would have been difficult and very expensive. Eventually they accepted computations compiled by ferrocement boat builders. In addition, we played it safe and built the studio stronger than a boat's hull by using more mortar and a stronger metal frame."

Except for the foundation, the method of construction paralleled the building of a ferrocement boat. First, workers dug a 24″ trench for the foundation footing and wired in more than the required amount of rebar. Before pouring the footing, vertical six-foot sections of 1″ rebar were attached to the foundation's rebar. Sticking out of the buried foundation, these vertical pieces served as the starting place for the thousands of feet of metal that went into the frame. Except for the door frames, made of 3″ metal I-beams, the metal consisted of ³⁄₈″ and ½″ rebar.

Workers spent hundreds of hours bending and shaping strands of rebar to make the proportions of the emerging metal skeleton conform to the scale model. Strands were strung vertically, then horizontally (forming rectangles about 10″ x 12″ at the bottom; smaller towards the top), and finally diagonally. At each junction or crossing, the rebar strands were wired together. After the finished frame—resembling a birdcage big enough for a pterodactyl—was in place, workers spent scores of hours wiring metal lath mesh to the outside of the frame.

Ferrocement is a poor insulator. In cold regions, ferrocementers build a second cage about six inches inside the first and insulate between them. Others have experimented with spraying insulation onto the walls. But in a climate

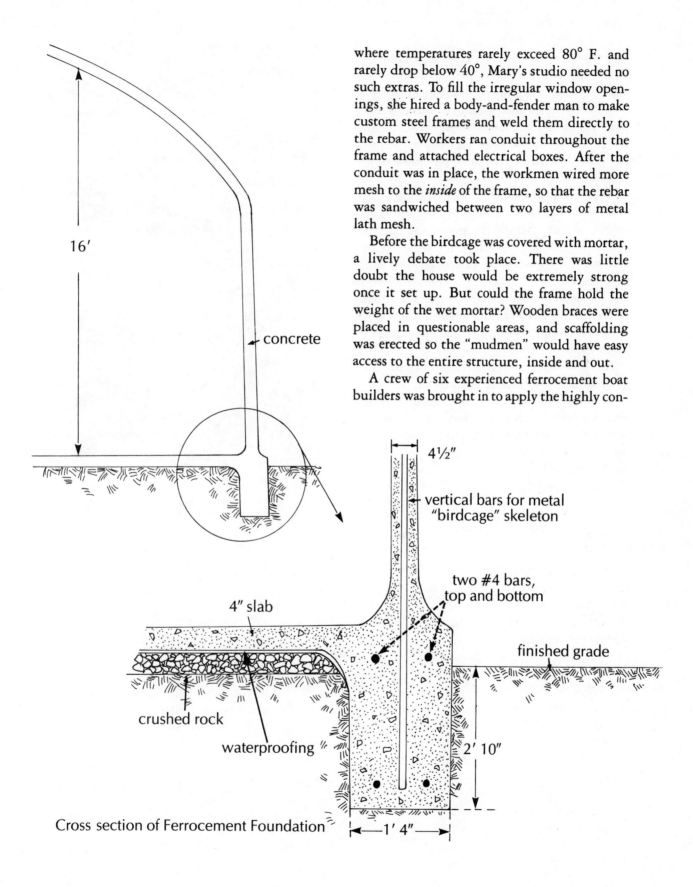

16'

concrete

where temperatures rarely exceed 80° F. and rarely drop below 40°, Mary's studio needed no such extras. To fill the irregular window openings, she hired a body-and-fender man to make custom steel frames and weld them directly to the rebar. Workers ran conduit throughout the frame and attached electrical boxes. After the conduit was in place, the workmen wired more mesh to the *inside* of the frame, so that the rebar was sandwiched between two layers of metal lath mesh.

Before the birdcage was covered with mortar, a lively debate took place. There was little doubt the house would be extremely strong once it set up. But could the frame hold the weight of the wet mortar? Wooden braces were placed in questionable areas, and scaffolding was erected so the "mudmen" would have easy access to the entire structure, inside and out.

A crew of six experienced ferrocement boat builders was brought in to apply the highly con-

4½"

vertical bars for metal "birdcage" skeleton

two #4 bars, top and bottom

finished grade

4" slab

crushed rock

waterproofing

2' 10"

Cross section of Ferrocement Foundation

1' 4"

centrated mortar mix: 2 parts clean sand to 1 part Type I-II Portland cement. Mary's crew started early and worked nonstop in order to avoid cold bonds that would crack and weaken the structure. The crew finished the first application in one day and came back to plaster inside and outside with a finer-grained mortar that left a smoother finish.

Workers mortared and tiled the bathroom spa. For downstairs flooring, Mary chose aggregate concrete. Before the pour, she had the plumbing put in, put down a plastic barrier, and laid ½" copper pipe for radiant floor heating. She painted the inside and outside white and installed carpeting in the upstairs bedroom. In the second story railing, a friend cleverly included a ladder that can be set up quickly to reach the uppermost periscope.

After nearly ten years, Mary's studio is in good health. No cracks have developed—not even around the fireplace, which experiences severe temperature changes. Mary: "If an economical material with the properties of plaster and superior insulation were developed, there would certainly be far-reaching effects in the building industry."

Second story railing

19.   *Houseboats*

IN ASIA, people get married, give birth, do business, and die on their boats. In Europe, families live on commercial diesel-driven barges, carrying tons of cargo, that chug up and down the waterways. In the United States, there are "river families" on the Mississippi and other large rivers, a relatively small group of roving yachtsmen, and sizeable houseboat communities in Fort Lauderdale, Seattle, and Sausalito.

Houseboats—expressions of individuality— offer a home for everybody's pocketbook and taste. Some are built for next to nothing, but others cost $300,000 and up. Many are built to code; others are not. There are almost as many styles as there are houseboats—Victorians, log

At anchor, houseboats display a variety of styles

Houseboat with detached "bedroom boat" with ferrocement foundation (bedroom floor is beneath the waterline)

cabins, A-frames, colonials, barn styles, but most employ a free-form design sheathed in natural woods.

The first houseboats appeared in Sausalito in the tolerant days when owning a small piece of San Francisco Bay shoreline didn't make a person a millionaire. Most were built on old hulls of oceangoing lifeboats, World War II landing craft, various kinds of barges, and anything roomy and able to float. By the late 1960s, houseboats gained greater acceptance. Middle-class professionals began building or buying them.

Calm water is the only unyielding criterion for a houseboat mooring. After that, feasibilty is contingent on local authorities, public sentiment, and initiative. Three principal kinds of flotation are available. The least expensive is encased Styrofoam, priced at $70 for each 4' x 3' x

7' section, secured to a metal or wood frame. Just as popular, but about twice as expensive, are sphere buoys, four feet in diameter, welded to a metal frame in the desired configuration. The most popular and expensive "foundation" is made of ferrocement, usually resembles a shoe box, and comes with an anchor-bolted redwood mudsill, just like a conventional house foundation. The most common dimensions are 18' x 30' and 5' deep.

Regardless of the choice of foundation, the rules are the same as on land: builders are supposed to follow the Uniform Building Code. But there is also an unwritten ethic that requires each boat to have an unconventional image or to make a statement. In a row of ten houseboats, typically, there will be ten defiantly different silhouettes.

In 1970, businessman and attorney Tom

Watson bought his typically atypical houseboat for $18,000 from the artist who built it. It sits on a 18' x 30' ferrocement hull. Each point of the 36' x 25' diamond-shaped living area cantilevers over the water. Inside, the end points of the diamond hold the bathroom and kitchen, and the middle serves as parlor, dining area, and small bedroom. A ladder leads to a small loft outfitted with a bed, skylight, and plenty of windows to enjoy the nightly glitter of millions of shore lights.

The walls, framed with standard 2" x 4"s, are insulated with fiberglass. On the inside, they are sheathed in good quality weathered barn wood, with 8" redwood tongue-and-groove siding on the outside. Tom's open-beamed ceiling is sheathed in pine. Plywood, rigid 1" insulation, and shingles complete the roof.

The electrical wiring and copper plumbing were installed in accordance with the UBC. As with all the other houseboats on Tom's pier, his sewage is collected in a 40-gallon drum that when full is pumped to the pier's sewage main. Because of the constant movement of the water, the drum is connected to the main with 4" flex pipe.

When Tom bought it, the houseboat was nicely built, but spartan. His first embellish-

Fireplace crafted from steel buoy

ment was the addition of several stained-glass windows. "I met some good glass artists and commissioned them to make windows over a period of time, so I could afford them." Tom also installed a fireplace made from a steel buoy. "It holds a fire overnight. I haven't used much gas since it was installed." Finally, Tom remodeled the kitchen and bathroom.

After ten years, Tom wanted more space. "I felt a little cramped. I had a good year financially, and I was at the stage in my life when I could build without worrying about costs. I wanted a super bedroom and bath that would make me happy the rest of my life. Since I didn't want to alter the lines of the houseboat, I decided to build a 'bedroomboat' on a welded-buoy foundation and attach it to the house with steps."

The people Tom hired to do the work were an interesting lot: "The two carpenters were former English teachers, the plumber was an unemployed physics teacher, and the electrician was a former biology teacher. They were perfectionists." Perfectionism, good taste, and money created an exquisite bedroom/bathroom. The bed sits on a curved raised platform of cherry wood and red spruce. The wall behind the bed is paneled in squares of Honduras mahogany, and the other walls are covered in kiln-dried cedar. A large double-paned skylight, plush wall-to-wall carpeting, and stained glass windows finish off the room.

The outside walls are covered with 8" redwood siding to match the main house. For roofing, Tom used Dex-o-Tex, which is installed in two steps: first, a porous fabric is

Bedroom interior

Paneled "basement," mostly below waterline

tacked to the roof and then a liquid is rolled onto the fabric. When the material sets up, it remains elastic and soft to the touch.

After finishing the bedroomboat, Tom sheathed the main houseboat's basement in pine. Most of the hull is below the waterline, so the water is only a foot below the basement windows.

Tom loves his houseboat: "Living on water has a tranquilizing effect. I love to come home and get absorbed in the natural rhythm of nature and let the day's activities shrink into proper perspective. A houseboat allows a person to stand aside from society and get spiritually recharged."

Houseboaters do live differently from their landlubber compatriots. They may awake to the scream of shore birds congregating to feed on a school of fish. They may be barked at by a hoarse harbor seal who comes by for a handout, or who has driven a school of anchovies under the boats. A houseboat is a vantage point on the pulsating elements that govern the natural world. There is a subtle but constant motion from the tides, currents, or from a storm that churns the deep water and sets the boats rocking and creaking in their moorings. A houseboat community soon develops its own character, politics, status, and attitudes towards outsiders. Issues like pirates (squatters), pier maintenance, sewage disposal systems, severe tides, anchor-outs, parking lot policy, and bird guano are common topics of conversation.

In 1966, Mimi Tellis anchored her old ferry boat/home in the calm waters of Richardson Bay. Mimi liked the Sausalito lifestyle, so she stayed. It became a drag to row her dinghy to shore for supplies, so she built a rickety, fifty-yard, pontoon-supported walkway from her ferry to shore. Soon her floating home was joined by other houseboats, some soundly and tastefully built, others resembling floating chicken coops. Mimi was happy to see kindred souls, but she was aware of the age-old maritime tradition of paying to stay at someone else's dock. She started charging rent.

Meanwhile, there was a snafu on land. The dock came ashore on someone's property, and naturally the owner felt uncomfortable with an access road emerging from the water. Mimi negotiated and ended up buying enough shoreline for a parking lot for her renters' cars, with an easement to the nearest public road.

About this time, the government officials became concerned about the floating subdivision that seemed to make a mockery of the ground-rules followed by other subdivisions. The boat owners and County officials reached mutually acceptable standards. Sewage from all units on Mimi's pier would be pumped into the city sewage system, and all houseboats would comply with local building codes.

A pier improvement and maintenance fund was established. Mimi replaced her walkway with a permanent pier of pile-driven poles and redwood decking. Later she added railings and attractive planters. Today she lives off her rentals, overseeing a pier with over thirty boats, and the long waiting list for berth openings speaks for a bright future.

Other piers have been established in Richardson Bay with rules and regulations worked out along the way. The fly in the ointment is a certain area of the waterfront, where

Skylight in "bedroom boat"

over a hundred people unwilling to play by the rules have ensconced themselves in mostly rent-free shacks and crude boats. Drug dealers, prostitutes, drunks are prevalent. The rule-abiding houseboaters and the surrounding community dislike this "outlaw" community, which brings unfavorable notoriety to the area.

When County officials attempted to cite and evict violators, they were pushed off the pier. Obviously a large force could remove the pirates and anchor-outs, but local officials are squeamish about a confrontation that might result in injury or loss of life. This bureaucratic frustration has brought down more repressive behavior towards all houseboaters, explains a legal houseboater whose home is worth $200,000: "We all live under constant scrutiny. Building inspectors patrol the area on weekends, presumably to cite flagrant violators.

They're too intimidated to write citations for real offenders, so instead of risking getting beat up, they walk around the civilized piers looking for infractions.

"A little-used portion of the code says any improvement valued above $200 must be approved by the Planning Department. So a neighbor was cited for installing a planter. You think anyone on *land* would be cited for installing a planter? Another neighbor was cited for paneling the inside of his basement; an inspector heard him hammering while sitting in the parking lot."

Angered by events affecting his community, one anonymous houseboater offers this advice: "If you build in a public place, expect to live by the rules. If you build in some hidden or remote waterway, *then* you can live by your own rules. You can't have it both ways."

20. Houses of Pounded Earth

To build houses, David Easton uses dirt. He scoops it up at the building site, mixes in a tad of cement, dumps it into wall-high forms clamped to a concrete foundation, and pounds it with pneumatic tampers until it becomes a rock-hard wall. He has combined his dirt medium with innovative construction techniques and new technology to build fully-equipped homes for 30 to 40% less than comparable conventional structures would cost.

A force to be reckoned with, David Easton wants to harness his Stanford University engineering degree and his high energy level to revolutionize housing construction in America: "There is an acute need for affordable housing. Frame construction, with its dependence on

David Easton mixing soil with cement

energy-intensive materials and dangerous synthetics, can no longer be relied upon to fill that need. Rammed earth will be right for the 21st century. Rammed earth walls are dry, fireproof, rot-proof, sound-proof, termite-proof, excellent passive solar collectors—and best of all, dirt cheap." To get out his message, David invites congressmen, governors, or anyone vaguely interested to tour one of his earthen houses. In his recently self-published book, *The Rammed Earth Experience* (Rammed Earth Works, Blue Mountain Road, Wilseyville, CA 95257), he unselfishly spells out all he has learned about rammed earth construction. The name of his construction and consulting firm

embodies David's double message: pounded dirt is his medium, and building with dirt really *does* work.

He knows he's onto something, because construction trade unions and corporate kingpins in the construction industry have lobbied against his low-cost housing proposals, complaining to government agencies that Easton's houses pose a threat to the industry. Of course, Easton points out that the industry wouldn't feel threatened if it were fulfilling the homebuyers' needs. "They lobby because they know they won't be able to compete when my idea catches on."

Actually, compacting earth to make walls is

Pounding dirt into the form

A new wall section

nothing new. Much of China's Great Wall is built of rammed earth. The ancient Assyrians, Babylonians, and Sumerians built their 150-foot ziggurats of rammed earth and mud brick. Rammed earth was first employed in California in the 1860s by the Chinese who worked the railroads; some of their buildings are still standing. During the Great Depression, architect Thomas Hibben supervised the creation of rammed earth housing in a federally-funded housing project in Alabama; many of *those* buildings still stand. David: "Some people conclude dirt construction is for backward cultures and poor people. That's wrong. When you combine earth and modern construction technology, the result is a superior product."

Easton found out about rammed earth from Ken Kern's book, *The Owner Builder*. He read everything else he could find, which wasn't much. "Something in me said, 'You have to try

it.'" In 1979, David sold his beautifully restored house on the California coast for a handsome profit and moved to Wilseyville, in the Sierra foothills. For $17,000, he tamped out his first house, which was just big enough for him, his wife Janis, and two small children, Kyber and Darth. In 1981 six other families commissioned David to build them rammed earth houses. During the process, he made peace with skeptical building departments in California, Kansas, and Colorado and developed an efficient, methodical approach to rammed earth construction.

David Easton's type of house is very well thought out: "You can think of it as two hands of God; one sweeps in and lifts up the sod roof, the other makes nice straight walls. A happy family moves into the earthen envelope, and the ecosystem is none the worst for it."

Choosing a site is important: "The site must

fit the building, not the other way around. Earthen walls absorb and store solar energy, so walls should make maximum use of winter sun. Any ground water must be diverted away from earthen walls. The site should be flat and have plenty of room to 1) build a house, 2) stockpile two large mounds of earth, and 3) maneuver a tractor to all the walls. If you can't lift the dirt into the forms with a tractor, you'll end up doing it by hand."

Easton and his workers scrape the site with a tractor. "The topsoil should be scraped away and piled separately for use later on the roof. It's never used in building. The building dirt is under the topsoil. As a rule, six inches scraped from the area the building's going to occupy provides the material for the walls. Six inches from a 30' x 40' building site will yield 22 cubic yards of material."

Once the pad has been leveled, Easton and company pour a sturdy concrete perimeter foundation. Easton designed a foundation with a ledge so the wall forms can be easily clamped to it.

Easton's wall forms are made of plywood, reinforced with horizontal trusses that help the plywood stand up to the heavy lateral forces exerted by the pneumatic tamper. The forms are essentially two 6' x 7' reinforced plywood sheets. Using pony or pipe clamps, Easton tightens the form to the ledge. Usually the walls are made 14" thick, but in colder climates, they can be made as thick as necessary. A form can be set up and ready for dirt in about five minutes, and taken down just as fast.

Ideally, the soil at the site should be 30% clay and 70% sand. With this proportion, proper wetting and tamping cause a chemical

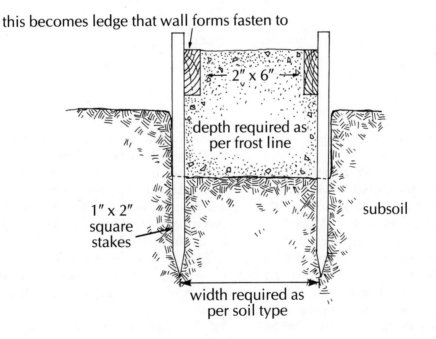

Cross Section of Concrete Foundation

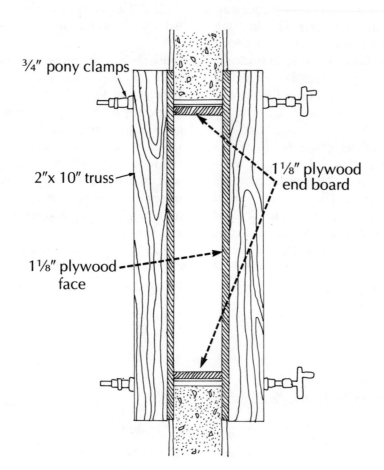

¾″ pony clamps

2″x 10″ truss

1⅛″ plywood
face

1⅛″ plywood
end board

change: colloids in the soil bond and become rockhard.

The soil composition is not always ideal, of course, and when it isn't, Easton makes adjustments. "If we have to add anything, it's usually sand, which isn't very expensive." When a customer requests it, Easton will have a soil engineer analyze the soil, but he's devised a less costly test: "Put two cups of soil into a quart jar. Fill the rest of the jar with water, and shake it until all the soil is in solution. After the jar has set overnight, there should be a visible line between the sand on the bottom and the clay on top." David warns not to use topsoil in the test. "It's full of organic matter that won't give you

the information you want. Remember, the building soil is *underneath* the topsoil."

Even with the preliminary tests, the compaction and bonding properties of soil aren't known until the first wall is made. "We always build an experimental wall first to see how it sets up. We adjust the mixture as necessary."

To prepare a pile of soil, Easton cranks up a rotary tiller and churns away. If the soil needs more sand, he mixes it in. Easton lives in earthquake country, so to the soil mix he adds about 5% Portland cement—which, as stress tests have shown, can add 500% more strength to the wall. As he works, a helper sprays the mix to make it damp, but not wet. Any roots, sticks, or large rocks are thrown out.

Scooping from the pile with a small tractor, he dumps about six inches into the form and tamps it. Easton has used a 4″ x 4″ x 6′ hand tamper and a pneumatic tamper that rents for $150 a day. Regardless of the tool, the test for complete compaction is not felt or seen but heard: "Tamp away. If it won't harden, it's too dry. If it's too rubbery, it has too much clay. If it's too dusty, it has too much sand. When it starts to compact, you'll hear a dull thud, then a sharp ringing sound. The sharp resonating means you've done it. Four inches of soil compacts into 2″ of wall."

During the pounding, Easton takes time out to stick the electrical boxes in the walls. The boxes are attached to a conduit that pokes out the top of the form. Easton puts a wedge-shaped wooden block above each box, and a block of wood inside it so that it isn't crushed in the tamping process. Later, wire is run down the conduit to the boxes. Where a door or window is called for, Easton adjusts the distance between panels or the height of tamped earth in a form.

In areas with no doors or windows, Easton leaves eight inches between the sections. Later these spaces will be wired with steel and poured

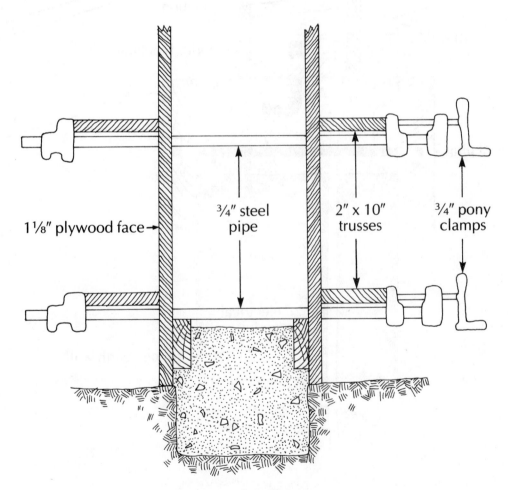

Side View of Wall Forms

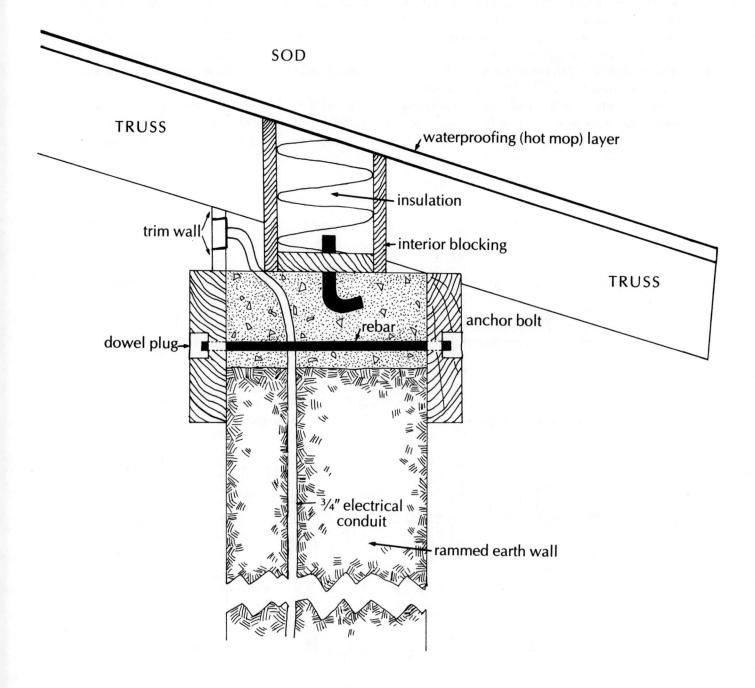

SOD

TRUSS

waterproofing (hot mop) layer

insulation

trim wall

interior blocking

TRUSS

anchor bolt

rebar

dowel plug

¾" electrical conduit

rammed earth wall

Working color into the cement floor

with concrete to form concrete posts that tie into a concrete bond beam that sits on top of the finished walls. Once all the wall sections are in place and the forms are pulled off, the home-to-be resembles Stonehenge. Easton ties the wall sections together with his "topside foundation." Attractive forms are built along the top of the wall in the eight-inch-wide spaces between sections. He pours concrete in the forms which also have rebar in them; taking precautions not to spill cement on the topside forms, because they will serve as permanent natural facing over the concrete. The bond beam, as the header for doors and windows and concrete posts, ties the walls together and prevent any movement from settling or earthquakes.

Easton grumbles when he talks about the topside foundation. It drives his construction costs up 25%, and he feels properly-built walls don't need that much reinforcing. To back up his statements, he cites comprehensive tests performed by the United States Bureau of Standards in 1940. But by using the topside foundation, his houses have passed muster with some of the most skeptical earthquake-conscious building departments in California. A wooden post-and-beam frame would be a less expensive way to tie the walls together, but some building departments fear a wood frame would flex too much in an earthquake.

To support the sod roofs, Easton always builds heavy-duty beam ceilings. A lighter roof

Floor stamping

with conventional synthetic insulation would be less expensive. But Easton, a purist, points to the carcinogenic properties of foam and fiberglass. He offsets costs by milling the wood himself and subcontracting out the tar roof. He spreads topsoil about 6″ deep on the tarred roof and then lays the sod. Wet, the roof must hold up fifty pounds per square foot.

For flooring, Easton uses either concrete or the soil mix used in the walls (but not as wet). He lays plumbing in the soil beneath the floor. If the floor is concrete, he uses dye to color it. As the slab hardens, workers press on the concrete or soil mix with a tool that stamps out 12″ squares. Once dry, the lines are grouted and the

floor is waxed, leaving a finished floor that looks just like tile.

Most walls don't require a sealer if the eaves keep runoff away. But walls that are rained upon directly will erode over a period of years. Easton has used three sealers: The easiest to apply is linseed oil mixed with turpentine. Walls can also be covered with stucco, but Easton prefers Dagga plaster, a mix of two parts fine sand, one part screened soil, and 10% plastic cement. For a heavy-duty sealer, extra thermal mass, and added decoration, Easton covers dirt walls with mortared brick.

"The scope of rammed earth is unlimited. Someday developers will have big dirt harvest-

Earth walls before
column and beam assembly has been formed

Interior of house, during construction

ing machines that will tamp out whole communities. Think of a machine that ingests undisturbed ground, saves the topsoil for the roofs, and pounds the walls out of what's underneath! There will be no dangerous synthetics packed into a house. No forests will have been cut down and shipped thousands of miles to building sites. The amount of BTUs that goes into building a house will be decreased 100 times."

Epilogue: Other Ideas

Think of a long, meandering road that can't be found on any map. Regulations and convention discourage visitors, but rumor has it that this road is lined with scores of innovative, affordable structures, each chock-full of fresh ideas. In good economic times, most people pass by the road without giving it a second thought; but in tough times, more and more people try to find it. It can be a frustrating search. Only dim silhouettes are visible, with incomplete impressions of unique structures.

Renegade Houses has profiled the first twenty buildings on this hypothetical road, allowing people to take a closer look. If you look on down the road past the houses discussed here, you'll notice other structures that have been converted from small schoolhouses, branch libraries, barns, stables, lighthouses, silos, detached residential garages, Dutch-style windmills, summer hideaways and vacation lodges, and train cars—cabooses, diners, boxcars, and sleepers. Anything with a roof over it is a potential house, and anything not commonly thought of as a house can often be purchased inexpensively. And it could be that not too far down the road is a house that's right for you.

About the Author

Eric Hoffman, born in 1946, is the grandson of W. D. Hoffman, the 1930s western novelist whose work was adapted to Tom Mix movies. Eric has taught English, History of Architecture, Anthropology, Journalism, and classes for teenagers with learning disabilities in the San Francisco Bay Area. During summer vacations, he took up home construction and in 1978, left teaching to pursue writing and building. In four years he's become an established magazine writer whose work covers a broad range of subjects.

Eric and his wife, Cecile Champagne, have been instrumental in introducing llamas to the American public as pets, pack animals, and wool sources. Eric's stories and experiences with llamas have appeared in *Adventure Travel, Monterey Life, Outside, California Today,* and a number of other publications. One story, "Sunny, the Pioneer Llama" helped win the National Headliner and Maggie Awards for *California Living.* The idea for *Renegade Houses* came from firsthand experiences while building a home on a mountaintop along the California Coast.

Bibliography

The following books, while not extensive in number, are in our opinion the best available on building and remodeling. Collectively, they make an excellent core library for the serious owner-builder.

I. *Housebuilding: Planning and Structure*

Alth, Max. *Do-It-Yourself Roofing and Siding.* New York: E. P. Dutton, 1978.
Clear explanations, with photos and drawings, of most roofing and siding materials and procedures.

Anderson, L. O. *How to Build A Wood-Frame House.* New York: Dover Publications, Inc., 1970. Reprint of a government classic, with extensive diagrams of most building steps.

Bingham, Bruce. *Ferro-Cement Design, Technique and Applications.* Centerville, MD: Cornell Maritime Press. Useful for boatbuilding.

Brunskill, R. W. *Illustrated Handbook of Vernacular Architecture.* New York: Universe Books, 1970. A portion of the book is devoted to wattle and daub.

Duncan, S. Blackwell. *How to Build Your Own Log Home and Cabin From Scratch.* Blue Ridge Summit, PA: TAB Books, 1978.

Easton, David. *The Rammed Earth Experience.* Wilseyville, CA: Blue Mountain Press, 1982.

Kern, Ken. *Owner Builder and the Code.* New York: Scribner Book Companies, Inc., 1977. Case studies of people's experiences in building their homes.

Kern, Ken. *The Owner-Built Home.* New York: Scribner Book Companies, Inc., 1975. Basic concepts of home design and how-to information on a variety of construction styles.

Love, T. W. *Construction Manual: Rough Carpentry.* Solona Beach, CA: Craftsman Book Company, 1976. Specific techniques for framing, sheathing, stairs, and insulation.

Morse, Edward S. *Japanese Homes and Their Surroundings.* New York: Dover Publications, Inc., 1961. Authentic study of traditional homes before Westernization.

Nervi, Luigi Pier. *Structures.* New York: McGraw-Hill Book Company, 1956. Has section on reinforced concrete.

Newcomb, Duane. *The Owner-Built Adobe House.* New York: Charles Scribner's Sons, 1980.

Roskind, Robert. *Before You Build: A Pre-Construction Guide.* Berkeley, CA: Ten Speed Press, 1981. A workbook to prepare you for all tasks required before building.

Sarviel, Ed. *Construction Estimation Reference Data.* Solona Beach, CA: Craftsman Book Company, 1981. A comprehensive reference book for material and cost estimating.

Wagner, Willis. *Modern Carpentry.* South Holland, IL: Goodheart-Wilcox Company, 1979. Used in apprenticeship courses, includes all phases of construction.

Williams, Benjamin. *Rafter Length Manual.* Solona Beach, CA: Craftsman Book Company, 1979. Planning and construction procedures for roofs; extensive tables.

II. *Remodeling: Planning, Repairs and Structure*

Hoffman, George. *How to Inspect a House.* New York: Delacorte Press, 1978. Guidelines for evaluating a home for remodeling or purchase.

Oakland Planning Dept. *Rehab Right*. Oakland, CA: Oakland Planning Dept. Considers design and *renovation* elements in remodeling a pre-1950 home.

Reader's Digest Editorial Staff. *Reader's Digest Complete Do-It-Yourself Manual*. New York: Reader's Digest Press. A comprehensive encyclopedia of common household repairs.

Williams, Benjamin (*ed.*). *Remodeler's Handbook*. Solona Beach, CA: Craftsman Book Company, 1976. Covers major remodeling projects, with planning and construction information.

Williams, T. Jeff. *All About Basic Home Repairs*. San Francisco, CA: Ortho Books, 1980. Clear, complete instructions.

III. *Finish Work: Housebuilding and Remodeling*

Foley, Joseph. *Electrical Wiring Fundamentals*. Intended for professionals, covers theory of electricity and all aspects of residential wiring.

Love, T. W. *Construction Manual: Finish Carpentry*. Solona Beach, CA: Craftsman Book Company, 1974. Detailed information and diagrams: roofing, siding, drywall, windows, stairs and trim.

Love, T. W. *Stair Builder's Handbook*. Solona Beach, CA: Craftsman Book Company, 1974. A book of tables for calculating rise, run, and headroom for stairs.

Massey, Howard. *Basic Plumbing With Illustration*. Solona Beach, CA: Craftsman Book Company, 1980. Covers planning and installing plumbing systems, with tables and diagrams.

Ortho Books. *Basic Wiring Techniques*. San Francisco, CA: Ortho Books. Clear, concise guide for home wiring projects from fixing a lamp to installing new circuits.

Ortho Books. *How to Design and Build Decks and Patios*. San Francisco, CA: Ortho Books. Ideas, construction details and directions for wood decks and masonry.

Rural Industries Bureau. *The Thatcher's Craft*. London: Rural Industries Bureau, 1960. Includes section on thatched roof for wattle & daub.

Sunset Books, *Remodeling with Tile*. New York: E. P. Dutton. Design ideas and installation tips; covers most types of tile.

IV. *Solar and Energy Efficiency*

Anderson, Bruce and Wells, Malcolm. *Passive Solar Energy*. Andover, MA: Brick House Publishing Co., 1981. A clear discussion of passive-solar theories with photos showing their application to various home designs.

Carter, Joe (*ed.*). *Solarizing Your Present Home*. Emmaus, PA: Rodale Press, 1981. Design principles and realistic projects using solar for existing homes.

Olkowski, Helga et al. *Integral Urban House: Self-Reliant Living in the City*. San Francisco, CA: Sierra Club Books, 1979. Solar, water and waste recycling, organic gardening, self-sufficiency.

Housing and Urban Development Dept. *Installation Guidelines for Solar Domestic Hot Water Systems*. Manual for installing solar hot water heating in a home.

Mazeria, Bruce. *The Passive Solar Energy Book*. Emmaus, PA: Rodale Press, 1979. A good source of information on passive solar design ideas and principles.

Shurcliff, William. *Thermal Shutters and Shades*. Andover, MA: Brick House Publishing Co., 1980. One hundred ways to reduce heat loss through windows.

Wing, Charles. *The Tighter House*. Analyzes cost-effective options for making a house more energy efficient.

For Further Information
OBC TRAINED ORGANIZATIONS

The Owner Builder Center (1824 Fourth St., Berkeley, CA 94710) trains various groups around the country to establish their own owner-builder schools.

BOMAR OWNER BUILDER CENTER
704 Gimghoul Road
Chapel Hill, NC 27514
Trained: 5/81, opened: 6/81
Robert Bacon and Marvin Pridgen,
(919) 929-3243

BUILDING RESOURCES
c/o Community Housing Investment Fund
121 Tremont
Hartford, CT 06105
Trained: 5/82, opened: 6/82
Dutch Walsh, (203) 233-5165

CENTRAL COAST OWNER BUILDER CENTER
1365 Nipomo Street
San Luis Obispo, CA 93401
Trained: 5/81, opened: 6/81
Bill Hood, (805) 543-8122

CHICO HOUSING IMPROVEMENT
PROGRAM
539 Flume Street
Chico, CA 95926
Trained: 5/82, opened: 6/82
Chris Lamb and Tom Weinberg,
(916) 342-0012

COLORADO OWNER BUILDER CENTER
1636 Pearl Street
Boulder, CO 80302
Trained: 2/81, opened: 7/81
Maureen McIntyre, (303) 449-6126

CORNERSTONES
54 Cumberland Street
Brunswick, ME 04011

FINE HOMES UNLIMITED
1461 Glenneyre, Suite E
Laguna Beach, CA 92651
Trained: 2/80, opened: 6/80
John Stebbins, (714) 494-9341

HANFORD ADULT SCHOOL
120 E. Grangeville Boulevard
Hanford, CA 93231
Trained: 2/81, opened: 7/81
Al Tolle, Principal,
Nick Schraa, Teacher, (209) 582-4401

HEARTWOOD OWNER BUILDER
CENTER
Johnson Road
Washington, MA 02135

MICHIGAN OWNER BUILDER CENTER
1505 East Eleven Mile Road
Royal Oak, MI 48067
Trained: 5/81, opened: 6/81
Ken Oxley, (313) 545-7033

MINNESOTA OWNER BUILDER CENTER
2615 Sixth Street, South
Minneapolis, MN 55454
Trained: 2/80, opened: 9/80
Richard Owings, (612) 339-5104

NORTHWEST OWNER BUILDER
CENTER
2121 First Avenue
Seattle, WA 98121
Trained: 2/81, opened: 6/81
Tom Phillips, (206) 447-9929

YESTERMORROW
Box 344
Warren, VT 05674
Trained: 2/80, opened: 6/80
John Connell, (802) 496-3437

Index